Rapid Assessment Program

A Rapid Biological Assessment of the Konashen Community Owned Conservation Area, Southern Guyana

Leeanne E. Alonso, Jennifer McCullough,
Piotr Naskrecki, Eustace Alexander and
Heather E. Wright (Editors)

RAP

Bulletin
of Biological
Assessment

51

Center for Applied Biodiversity Science
(CABS)

Conservation International

Conservation International – Guyana

The *RAP Bulletin of Biological Assessment* is published by:
Conservation International
Center for Applied Biodiversity Science
2011 Crystal Drive, Suite 500
Arlington, VA USA 22202
Tel : 703-341-2400
www.conservation.org
www.biodiversityscience.org

Editors: Leeanne E. Alonso, Jennifer McCullough, Piotr Naskrecki, Eustace Alexander and Heather E. Wright
Design: Glenda Fabregas
Map: Mark Denil
Photographs: Piotr Naskrecki (except where noted)

RAP Bulletin of Biological Assessment Series Editors:
Jennifer McCullough and Leeanne E. Alonso

ISBN # 978-1-934151-24-2

Suggested citation:

Alonso, L.E., J. McCullough, P. Naskrecki, E. Alexander and H.E. Wright. 2008. A rapid biological assessment of the Konashen Community Owned Conservation Area, Southern Guyana. RAP Bulletin of Biological Assessment 51. Conservation International, Arlington, VA, USA.

This RAP survey was made possible by generous financial support from the Leon and Toby Cooperman Family Foundation. Publication of this RAP Bulletin was generously funded by the Gordon and Betty Moore Foundation.

Table of Contents

Chapters

Appendices

Participants and Authors

Eustace Alexander (coordination, editor)
Conservation Science Manager
Conservation International-Guyana
266 Forshaw St., Queenstown
Georgetown, Guyana
ealexander@conservation.org

Leeanne E. Alonso (reconnaissance, editor)
Rapid Assessment Program
Center for Applied Biodiversity Science
Conservation International
2011 Crystal Drive, Suite 500
Arlington, VA USA 22202
l.alonso@conservation.org

Vitus Antone (videography, coordination)
Field Program Assistant
Conservation International-Guyana
Upper Takatu / Upper Essequibo
Region 9
Lethem, Guyana
vantone@conservation.org

Curtis Bernard (coordination)
Biodiversity Analyst
Conservation International-Guyana
266 Forshaw St., Queenstown
Georgetown, Guyana
cbernard@conservation.org

Andrew DeMetro (indigenous relations, coordination)
Indigenous Relations Advisor
Conservation International-Guyana
Upper Takatu / Upper Essequibo
Region 9
Lethem, Guyana
ademetro@conservation.org

Michelle Kalamandeen (field assistance)
Centre for the Study of Biological Diversity
Department of Biology, Faculty of Natural Sciences
University of Guyana, Turkeyen Campus, ECD
Guyana, South America
michellek@bbgy.com

Carlos A. Lasso (fishes and crustaceans)
Fundación La Salle de Ciencia Naturales
Apartado 1930 Caracas 1010 A
Caracas, Venezuela
carlos.lasso@fundacionlasalle.org.ve

Celio Magalhaes (decapod crustaceans)
Instituto de Pesquisas do Amazonia
Caixa Postal 478, 69011-970 Manaus
Amazonas, Brasil
celiomag@inpa.gov.br

Christopher Marshall (beetles)
Oregon State University
Department of Zoology, Cordley 4082
Corvallis, Oregon USA 97331
christopher.marshall@oregonstate.edu

Jennifer McCullough (editor)
Rapid Assessment Program
Center for Applied Biodiversity Science
Conservation International
2011 Crystal Drive, Suite 500
Arlington, VA USA 22202
j.mccullough@conservation.org

Lina Mesa (fishes)
Fundación La Salle de Ciencia Naturales
Museo de Historia Natural La Salle
Apartado 1930 Caracas 1010 A
Caracas, Venezuela
lmesasalazar@gmail.com

Julián Mora-Day (fishes and aquatic invertebrates)
Fundación La Salle de Ciencia Naturales
Museo de Historia Natural La Salle
Apartado 1930 Caracas 1010 A
Caracas, Venezuela
julianmoraday@gmail.com

Piotr Naskrecki (katydids and grasshoppers, editor)
Invertebrate Diversity Initiative
Center for Applied Biodiversity Science
Conservation International
2011 Crystal Drive, Suite 500
Arlington, VA USA 22202
p.naskrecki@conservation.org

Brian J. O'Shea (birds)
Louisiana State University Museum of Natural Science
119 Foster Hall – LSU
Baton Rouge, LA USA 70803
boshea2@lsu.edu

Gilson Rivas (reptiles and amphibians)
Museo de Historia Natural La Salle
Apartado 1930 Caracas 1010 A
Caracas, Venezuela
anolis30@hotmail.com

James G. Sanderson (large mammals)
Center for Applied Biodiversity Science
Conservation International
2011 Crystal Drive, Suite 500
Arlington, VA USA 22202
gato_andino@yahoo.com

Ted R. Schultz (ants)
Department of Entomology
Smithsonian Institution
POB 37012
NHB, CE516, MRC 188
Washington, DC USA 20013-7012
schultzt@si.edu

Josefa Celsa Señaris (reptiles and amphibians)
Fundación La Salle de Ciencia Naturales
Apartado 1930 Caracas 1010 A
Caracas, Venezuela
josefa.senaris@fundacionlasalle.org.ve

Hector Samudio (fishes)
Fundación La Salle de Ciencia Naturales
Museo de Historia Natural La Salle
Apartado 1930 Caracas 1010 A
Caracas, Venezuela
hectorsagaster@gmail.com

Jeffrey Sosa-Calvo (ants)
Department of Entomology
Smithsonian Institution
POB 37012
NHB, CE516, MRC 188
Washington, DC USA 20013-7012
sossajef@si.edu

Maya Trotz (water quality)
University of South Florida
4202 E. Fowler Ave.
ENB 118
Tampa, FL USA 33620
matrotz@eng.usf.edu

Heather E. Wright (coordination, editor)
Rapid Assessment Program
Center for Applied Biodiversity Science
Conservation International
2011 Crystal Drive, Suite 500
Arlington, VA USA 22202
Heather.Wright@moore.org

Wai-Wai Parabiologists

Bemner Isaacs
Konashen Indigenous District
Guyana

Elisha Marawanaru
Konashen Indigenous District
Guyana

Anthony Shu-Shu
Konashen Indigenous District
Guyana

Romel Shoni
Konashen Indigenous District
Guyana

Reuben Yaynochi
Konashen Indigenous District
Guyana

Charakura Yukuma
Masakenari Village, Konashen Indigenous District
Guyana

Organizational Profiles

CENTER FOR APPLIED BIODIVERSITY SCIENCE (CABS)

The mission of the Center for Applied Biodiversity Science (CABS) is to strengthen the ability of Conservation International and other institutions to identify and respond to elements that threaten the earth's biological diversity. CABS collaborates with universities, research centers, multilateral government and non-governmental organizations to address the urgent global-scale concerns of conservation science. CABS researchers are using state-of-the-art technology to collect data, consult with other experts around the world, and disseminate results. In this way, CABS research is an early warning system that identifies the most threatened regions before they are destroyed. In addition, CABS provides tools and resources to scientists and decisions-makers that help them make informed choices about how best to protect the hotspots.

Conservation International
2011 Crystal Drive, Suite 500
Arlington, VA USA 22202
www.biodiversityscience.org

CONSERVATION INTERNATIONAL

Conservation International (CI) is an international, nonprofit organization based in Arlington, Virginia, USA. CI believes that the Earth's natural heritage must be maintained if future generations are to thrive spiritually, culturally and economically. Our mission is to conserve the Earth's living heritage, our global biodiversity, and to demonstrate that human societies are able to live harmoniously with nature.

Conservation International
2011 Crystal Drive, Suite 500
Arlington, VA USA 22202
703-341-2400
www.conservation.org

CONSERVATION INTERNATIONAL – GUYANA

Conservation International has been active in Guyana since 1990 and operates under a Memorandum of Understanding (MOU) signed with the Government of Guyana. CI–Guyana's staff is comprised of 100% Guyanese, a group committed to assisting with the development of Guyana in an environmentally sustainable, culturally appropriate, and economically sensitive manner. The vision of CI–Guyana is to establish Biodiversity Corridors in Guyana, incorporating the anchors of a National Protected Area System, while developing trans-boundary corridors across the Guayana Shield. CI–Guyana has been working with the Government of Guyana and resident indigenous communities to develop long-term and sustainable management for the establishment of protected areas, helping to develop protected area legislation in Guyana and actively working with the Government to endow and establish a protected areas trust fund to finance the management of Guyana's protected areas in perpetuity.

Conservation International-Guyana
266 Forshaw St., Queenstown
Georgetown, Guyana

Acknowledgements

The success of this RAP survey would not have been possible without the collective effort of many dedicated individuals and organizations. First of all, we thank the Honorable Carolyn Rodrigues of the Ministry of Amerindian Affairs and Dr. Indarjit Ramdass of the Environmental Protection Agency of Guyana for permitting us to conduct research in the Konashen Indigenous District of Guyana (Research Permit No. 021006 BR 60). We are especially grateful for the collaboration and support from Wai-Wai community leadership, Touchou Cemci (James) Suse and elders of the Masakenari village. We are grateful to Celvin Bernard, Philip de Silva, and Michelle Kalamandeen of the Center for the Study of Biological Diversity at the University of Guyana for their help in obtaining necessary exportation permits for collected specimens. Special thanks to David Clarke for sharing his notes and knowledge of the flora of the COCA which helped us to select the RAP survey sites.

We thank the staff of CI-Guyana for their invaluable help in organizing and conducting the survey, especially Eustace Alexander, Curtis Bernard, Vitus Antone, and Andrew DeMetro. Local Wai-Wai parabiologists were absolutely indispensable during our time in the field, and we express our gratitude to Bemner Isaacs, Anthony Shu-Shu, Romel Shoni, Reuben Yaynochi, Elisha Marawanaru, and Charakura Yukuma. Their hard work, dedication, in-depth knowledge of the inner workings of boat motors and building camp sites, occasional provision of fresh food, and their inspiring companionship, helped make this expedition a success. Special thanks to our cook, Feon, who kept us nourished and well fed. We also thank Raquel Toniolo-Finger of the BBC Natural History Unit for her companionship, and badly needed fresh food supplies brought to our camp.

The RAP participants thank Conservation International's RAP program for the invitation to participate in this RAP survey. The editors thank Mark Denil of CI's Conservation Mapping Program and both Glenda Fabregas and Kim Meek for their attention to detail and patience in designing RAP publications.

The mammal team wishes to thank the Wai-Wai community leadership, especially Touchou James Suse, as well as Henry James, Elvis Joseph, Elisha Mauruwanaru, Romel Shoni, Antoni Shushu, Bemner Suse, Philip Suse, Rueben Yamochi, Christopher Marshall, Piotr Naskrecki, Brian O'Shea, Jeffrey Sosa-Calvo, Ted Schultz, Celsa Señaris, and Carlos Lasso for aiding the survey. The katydid team would like to thank Elisha Marawanaru for his indispensable help and keen eyes during night collecting expeditions.

We thank the Leon and Toby Cooperman Family Foundation for their generous financial support of this RAP survey. We also thank The Gordon and Betty Moore Foundation for funding publication of this RAP bulletin.

Report at a Glance

A RAPID BIOLOGICAL ASSESSMENT OF THE KONASHEN COMMUNITY OWNED CONSERVATION AREA, SOUTHERN GUYANA

Dates of RAP Survey

October 6–28, 2006

Description of RAP Survey Sites

The Konashen Community Owned Conservation Area (COCA) is comprised of 625,000 hectares of undisturbed forest located in a tropical wilderness area in the "deep" southern region of Guyana. The site is relatively unexplored and considered to be one of the last large and intact pristine areas of forest remaining in Guyana. It encompasses the watershed of the Essequibo River and the tributaries of the Kassikaityu, Kamoa, Sipu and Chodikar rivers. The area's main mountains include the Wassarai, Yashore, Kamoa and Kaiawakua with elevations reaching 1200 meters above mean sea level. Within the Konashen COCA, the RAP team surveyed two primary sites and the aquatic teams surveyed focal areas encompassing the main waterways.

Reasons for the RAP Survey

Within the Konashen COCA there is one community (Masakenari) made up mainly of members of the Wai-Wai Indigenous group who utilize the area for their sustenance. The residents of this village have minimal external contacts and thus utilize the natural resources of the COCA for all their needs. Their major form of economic activity is the international wildlife trade, but they may also harvest other raw materials from the forest to support a craft industry in the community. The Wai-Wai of the COCA recognize that their demands on their natural resources are increasing and must be managed sustainably. Therefore, they expressed interest in collaborating with Conservation International (CI) and the RAP program to conduct an inventory of the natural resources of the COCA. The data collected will be used by the community to establish user-thresholds and to develop a management plan for sustainable use and conservation of their traditional resources.

MAJOR RESULTS

The data collected during the RAP survey indicate that the forests of the Konashen COCA are in very good condition and support rich biodiversity. Water quality was high, with no evidence of pollution. Typical of the forests of the Guayana Shield, the RAP team recorded high species diversity but low abundance levels of species of most groups and low species endemicity. The potential for finding new taxa is high due to the lack of scientific exploration.

Number of Species Recorded

Ants	200+ species
Beetles (Scarabinae)	50+ species
Katydids	73 species
Fishes	113 species
Amphibians	26 species
Reptiles	34 species
Birds	319 species
Mammals	42 species (21 confirmed)

Species Possibly New to Science

Ants	at least 1 species (*Trachymyrmex* sp.)
Beetles	at least 1 species
Katydids	at least 7 species
Fishes	*Hoplias* sp.
	Ancistrus sp.
	Rivulus sp.
	Bujurquina sp.

New Records for Guyana

Katydids	58 species
Reptiles	*Typhlophis ayarzaguenai* (blind snake)
Birds	*Ramphotrigon megacephalum* (Large-headed Flatbill)
Ants	At least 1 species (*Mycetarotes acutus*)

Species of Conservation Concern (IUCN 2008 and CITES 2008)

Brown-bearded saki monkey (*Chiropotes satanas*), Endangered
Giant otter (*Pteronura brasiliensis*), Vulnerable
Giant armadillo (*Priodontes maximus*), Vulnerable
Bush dog (*Speothos venaticus*), Vulnerable
Brazilian tapir (*Tapirus terrestris*), Vulnerable
Chelonoidi carbonaria, Vulnerable
Blue-cheeked Parrot (*Amazona dufresniana*), Near Threatened
Scarlet Macaw (*Ara macao*), CITES Appendix I
Chelonoidis spp., CITES Appendix II
Black caiman (*Melanosuchus niger*), CITES Appendix I
Dwarf caiman (*Paleosuchus trigonatus*), CITES Appendix II
Emerald tree boa (*Corallus caninus*), CITES Appendix II

Species Endemic to the Guayana Shield
32 Bird species:

Black Curassow (*Crax alector*)
Caica Parrot (*Gypopsitta caica*)
Blue-cheeked Parrot (*Amazona dufresniana*)
Rufous-winged Ground-Cuckoo (*Neomorphus rufipennis*)
Guianan Puffbird (*Notharchus macrorhynchos*)
Black Nunbird (*Monasa atra*)
Guianan Toucanet (*Selenidera piperivora*)
Green Araçari (*Pteroglossus viridis*)
Golden-collared Woodpecker (*Veniliornis cassini*)
Chestnut-rumped Woodcreeper (*Xiphorhynchus pardalotus*)
Black-throated Antshrike (*Frederickena viridis*)
Band-tailed Antshrike (*Sakesphorus melanothorax*)

Northern Slaty-Antshrike (*Thamnophilus punctatus*)
Guianan Streaked-Antwren (*Myrmotherula surinamensis*)
Rufous-bellied Antwren (*Myrmotherula guttata*)
Brown-bellied Antwren (*Epinecrophylla gutturalis*)
Todd's Antwren (*Herpsilochmus stictocephalus*)
Guianan Warbling-Antbird (*Hypocnemis cantator*)
Black-headed Antbird (*Percnostola rufifrons*)
Ferruginous-backed Antbird (*Myrmeciza ferruginea*)
Rufous-throated Antbird (*Gymnopithys rufigula*)
Boat-billed Tody-Tyrant (*Hemitriccus josephinae*)
Painted Tody-Flycatcher (*Todirostrum pictum*)
Capuchinbird (*Perissocephalus tricolor*)
Guianan Red-Cotinga (*Phoenicercus carnifex*)
Guianan Cock-of-the-Rock (*Rupicola rupicola*)
White-throated Manakin (*Corapipo gutturalis*)
White-fronted Manakin (*Lepidothrix serena*)
Tiny Tyrant-Manakin (*Tyranneutes virescens*)
Tepui Greenlet (*Hylophilus sclateri*)
Blue-backed Tanager (*Cyanicterus cyanicterus*)
Golden-sided Euphonia (*Euphonia cayennensis*)

CONSERVATION CONCLUSIONS FROM THE RAP SURVEY

(see Executive Summary for more details)

The results of the RAP survey clearly support the Wai-Wai's decision to manage their area for conservation. To achieve their objectives of effectively managing the Konashen COCA, the residents of Masakenari need to address the perceived threats (illegal mining, logging, trapping, etc.) presented by the development of roads in neighboring areas. This will require them to establish community-based regulations and implement a system of vigilance and patrols to repel encroachment by illegal miners, loggers, and trappers from the area.

Furthermore, in order to sustainably manage their resources, the Wai-Wai will need to conduct further studies, including baseline inventories of species most likely to be over-harvested, and establish sustainable thresholds of hunting/collecting/gathering. Due to the socio-economic and cultural importance of hunting to the community, it is important for them to develop and implement a rotation system to distribute the effects of subsistence hunting over as large an area as possible.

In order to sustain their health, well-being and livelihoods while protecting their biodiversity, the residents of the area need to be regularly informed of any changes in the biological and ecological conditions of the COCA (e.g. the quality of their water resources and the status of populations of fish, mammals, vegetation, birds, etc.). This will require them to develop and implement a plan to monitor and assess the species populations that are the Wai-Wai community's most valuable food, hunting and trade resource, and also to establish a water quality monitoring program of the major rivers of the COCA.

Since the Wai-Wai's overarching goal of managing their COCA is to "keep their biodiversity" they need to continue to avoid the trapping and trading of species such as parrots and macaws for the pet trade, and to restrain from hunting and harvesting threatened species.

Ecotourism appears to be a viable conservation-based enterprise for the area. As such, the Wai-Wai community should implement a basic eco-tourism infrastructure that supports research and education-based activities that will further enhance conservation efforts and biological knowledge of the COCA.

As the Wai-Wai's management of the COCA progresses, there will be a need to further enhance the human and technical capacity for management. This will require the formation of formal partnerships with training, research and other conservation institutions, as well as the continued development of the Wai-Wai Rangers to effectively implement actions for the conservation management of the aquatic and terrestrial resources of the COCA.

Executive Summary

INTRODUCTION

The Guayana Shield Region

The Guayana Shield Region of northern South America (approx. 8⁰N, 72⁰W) was formed during the Precambrian era and is one of the most ancient landscapes in the world. The terms Guiana and Guayana are two universally accepted variants of an Amerindian word interpreted to mean "land of plenty water." Participants of the Guayana Shield Priority Setting Workshop (Huber and Foster 2003) identified the Guayana Shield as the area bounded by the Amazon River to the south, the Japura-Caqueta River to the southwest, the Sierra de Chiribiquere to the west, the Orinoco and Guaviare rivers to the northwest and north, and the Atlantic Ocean to the east. This area covers 2.5 million km² of mountains, pristine forests, wetlands and savannahs, and is comprised of parts of Venezuela, Brazil and Colombia and all of Guyana, Suriname and French Guiana. The Guayana Shield occupies approximately 13% of the entire South American continent (Hammond 2005). Recognized as one of the world's largest remaining areas of tropical wilderness, the Guayana Shield Region is also home to a wide variety of unique ecosystems (e.g. tepui or table-top mountains), harbors a large number of endemic fauna and flora, and supports a high level of cultural diversity with more than 100 indigenous ethnic groups, most of whose cultures remain relatively unblemished and are intimately dependent upon the natural resources of the region for their sustenance.

Guyana

Guyana is located on the northern coast of South America and is bordered by the Atlantic Ocean to the north, Suriname to the east, Venezuela to the west and Brazil to the south and southwest. Several features distinguish Guyana's forests from other regions of the world. A primary distinction is that 80% of the country is forested and 75% of this remains relatively intact. This is one of the highest percentages of pristine tropical habitat for any country. Other distinguishable features include the underlying geologically old Guayana Shield, close proximity to the biologically rich Amazon Basin, low population density and an interior that is relatively inaccessible. These factors have contributed to Guyana's richness in biodiversity and biologically important habitats.

The temperature is tropical with average high daily temperatures of 25.9 ˚C and average high rainfall of 4400 mm per year. The major geographic regions include the narrow coastal strip (occupied by more than 75% of the national population) which is basically a floodplain that lies approximately 2 m below sea level at high tide and is dissected by estuaries of 16 rivers, streams, creeks and canals for drainage and irrigation, the hilly white sandy region lying just behind the coastal plain, the savannahs in the central and south-western parts of the country, and the highland regions of the Acarai, Imataka, Kanuku and Pakaraima mountain ranges.

Guyana's large expanses of freshwater ecosystems include the Essequibo River – the third largest water source in South America. The numerous waterways intermix annually with tributaries of the Amazon River (for example, the Rio Negro and Rio Branco River) during the rainy season when the banks overflow and flood the Rupununi Savannahs. This connectivity facilitates trans-boundary migration of biodiversity – especially species seeking refuge from the more impacted places such as Roraima State in neighboring Brazil.

Additionally, Guyana has a wealth of species with many yet to be discovered and recorded. To date, knowledge of the country's biodiversity includes more than 7,000 species of plants

(Funk et al. 2007), almost 800 species of birds (Braun et al. 2000), 225 species of mammals (Engstrom and Lim 2008), and about 320 species of reptiles and amphibians (Hollowell and Reynolds 2005). Many of these species are endemic, rare or vulnerable locally, and/or threatened in other parts of the world. For example, 173 plant species are considered endemic and 17 faunal species are already rare or threatened (IUCN 2008). The threatened species found in Guyana include the Black Caiman (*Melanosuchus niger*), Harpy Eagle (*Harpia harpyja*), Arapaima (*Arapaima gigas*), Giant River Turtle (*Podocnemis expansa*), the Giant River Otter (*Pteronura brasiliensis*), and large predators such as the Jaguar (*Panthera onca*) (IUCN 2008). Although the interior is relatively inaccessible and contains a low population density, there are a few current and perceived threats to the biodiversity in the country. Inherently rich in natural resources, Guyana is traditionally dependent upon extraction industries. Over the last 15 years the Government of Guyana (GOG) has been issuing licenses to many large local and multi-national logging and mining companies for concessions in the interior. Small-scale mining and logging are also occurring in various parts of the interior.

Because the cash economy within the indigenous communities is not well developed, employment opportunities are limited and poverty is widespread. Consequently, the residents are inclined to trade in wildlife and fish to supplement their income. Guyana's national borders with neighboring countries are open and unmonitored. This allows easy access by other nationals into the country to illegally extract and/or trade in natural resources. The bridging of the Takutu River and the development of a road linking Georgetown with Brazil and other parts of South America via the Trans-Amazonia Highway are expected to exacerbate this threat.

To mitigate against these threats and protect the country's invaluable biodiversity, the GOG has declared its commitment to the development of a national protected areas system and is currently coordinating mechanisms for its implementation. To date, five areas have been identified for protection, one of which is in the Southern Region where the Konashen Indigenous District is located. As Guyana develops its system of protected areas, Conservation International (CI) has been identified by the GOG as the lead agency in the process. CI is therefore collaborating with the GOG and building partnerships with local stakeholders to develop protected areas in the Kanuku Mountains and Southern Guyana Region. Other organizations including the World Wildlife Fund, the Guyana Marine Turtle Conservation Society and Flora and Fauna International are also participating in the process of developing protected areas in other priority sites. The Iwokrama Centre and the Kaieteur National Park Board have the responsibility for developing and overseeing the implementation of management plans for the Iwokrama Forest and the Kaieteur National Park respectively.

The Konashen Community Owned Conservation Area (COCA)

The Konashen Community Owned Conservation Area (COCA) lies within the Konashen Indigenous District ($1^011'$ to $2^02'$N and $58^018'$ to $59^039'$W) in the tropical wilderness area of remote southern Guyana. The Konashen COCA is 625,000 hectares of pristine rainforests and is considered by many to be the last of the pristine frontier rainforests remaining in Guyana. It encompasses the watershed of the Essequibo River (Guyana's major water source) and the tributaries of the Kassikaityu, Kamoa, Sipu and Chodikar rivers. The site contains the Wassarai, Yahore, Komoa and Kaiawakua mountains with elevations as high as 1200 m a.s.l. The pristine state of the area is due its extremely low population density (about 0.032 humans/km²) and the difficult terrain, which negatively affects the accessibility and economic viability of potential extractive industries. Only one community – Masakenari – inhabited by fewer than 300 people, mainly of Wai-Wai ancestry, exists in the area.

Most of the forests in the Konashen COCA are tall, evergreen hill-land and lower montane forests, with large expanses of flooded forest along major rivers. Except for the flora, the biodiversity of the site is poorly understood. Challenges in accessibility are the main reason for lack of knowledge on the area. The Smithsonian Institution has identified nearly 2,700 species of plants from this region, representing 239 distinct families (Hollowell et al. 2001).

The forests of the Konashen COCA protect the southern watersheds of the headwaters of the Essequibo River, which forms part of the northern Amazon ecosystem. Hence the area is of great importance to the provision of quality freshwater to downstream communities and the entire country as a whole. Of great importance is the potential role of the Konashen COCA to the formation of a southern Guyana biodiversity corridor along the Essequibo River linking with the proposed one million hectare Conservation Concession expansion, North Rupununi Wetlands, the Kanuku Moutains Protected Area, the Iwokrama Rainforest Reserve and the Kaeiteur National Park. Konashen is also of strategic importance to the long-term vision of Conservation International, which is to link this southern Guyana corridor with other protected areas in the region to create a mega-Guayana Shield Tropical Wilderness Corridor.

In February 2004 the GOG issued to the residents of the Konashen Indigenous District an Absolute Title to their lands making them the legal guardians of the area. In order to mitigate perceived threats to their culture and resources, the community made a decision to manage their lands for biodiversity conservation and economic development. Recognizing that they lack the required skills and other forms of capacity for conservation management, they sought and gained the support of the both the GOG and Conservation International-Guyana (CIG) to develop a sustainable plan for their lands. In November 2004 the three parties signed a Memorandum of Cooperation (MOC), which outlined a plan for sustainable use of Konashen COCA's biological resources. In the MOC, the Wai-Wai of the Konashen Indigenous District and CIG specifically agreed to cooperate as follows:

- to jointly evaluate the ongoing resource needs of the Wai-Wai and the impact of traditional land uses on biodiversity and ecosystems.

- to jointly conduct surveys and other activities necessary to collect data for an adequate evaluation.

- to work together to increase local, national and global awareness of the importance of biodiversity and ecosystems on the Wai-Wai land.

- to jointly develop land-use practices that satisfy Wai-Wai needs while also preserving ecosystems and biodiversity.

- to develop an appropriate strategy for managing resource use and for identifying and addressing threats to the integrity of the area.

- to identify and formulate income-generating projects and potential sources of funding of the same.

- to work together to establish the Wai-Wai lands as a Wai-Wai owned and managed conservation area for future recognition and incorporation by the national protected area system.

- to work together to identify and secure adequate funds to finance the implementation of this collaborative process.

- to regularly collaborate to update the GOG, through the Ministry of Amerindian Affairs, regarding the implementation of the process, in order to benefit from its insight and contribution.

For more than seven years, CIG has been working with the community of Masakenari to build local capacity for assessment, monitoring and managing of their natural resources. Prior to the RAP survey, CIG trained six members of the local Wai-Wai community to survey the species richness of fishes, mammals and birds. The training included techniques to enumerate mammals through the use of camera phototraps, surveying fishes with nets, and the identification of birds. The primary focus of this training was on species that are captured and used by the Wai-Wai as food sources or in trade. This exercise created of a cadre of trained Wai-Wai community members at Masakenari who can now lead in the management of a subsequent long-term biological monitoring program for the Konashen COCA.

RAP Survey of the Konashen Community Owned Conservation Area (COCA)

Residents of the Konashen COCA depend upon the area for their sustenance and have recognized that their need for resources is increasing and, if not well managed, there may serious negative impacts in the future. Therefore, to ensure that there is sustainable utilization of their biological re-

sources, they expressed an interest to collaborate with CI to conduct a RAP survey of their local biodiversity.

The RAP data will be used to help the community to establish user-thresholds and allow for sustainable use of their traditional resources. The data collected will not only guide resource use but will also be used to guide zoning and management plans for the area which is being considered for legal protection status. An additional objective of the RAP survey was to build the Wai-Wai's capacity to facilitate their own data collection, analyses and presentation.

During October 6–28, 2006, CI's Rapid Assessment Program (RAP), in collaboration with CIG and the Smithsonian Institution, conducted a rapid biodiversity survey of selected sites in the Konashen COCA. The objectives of the RAP survey were 1) to collect baseline data on the biodiversity of the COCA for potential use in the development of a small-scale ecotourism industry managed by the Wai-Wai community; and 2) to use the information gathered on species used primarily for food and trade by the community to establish user-thresholds that ensure sustainable utilization of these resources. In addition, the RAP team assessed the water quality of rivers and streams in the vicinity of the community as well as the conditions of populations of animals hunted and fished by the Wai-Wai community.

The scientific team included scientists and students from the University of Guyana, the Smithsonian Institution, Conservation International, Oregon State University, Louisiana State University, University of South Florida and Fundación La Salle de Ciencias Naturales, Venezuela. Six Wai-Wai parabiologists assisted the scientists during the survey as a part of their forest ranger training. The team collected data on water quality and the following groups of animals: ants, katydids, dung and passalid beetles, decapod crustaceans, mollusks, fishes, amphibians, reptiles, birds, and large mammals. Most of data and specimen collecting took place at two main camps, the first located at the foothills of the Acarai Mountains along the Sipu River and the second located alongside Kamoa River, approximately 15 km east of the foothills of the Kamoa Mountains. The RAP team did not survey plants because plant collections had previously been made in the COCA by the Smithonian Instituion (D. Clarke, unpublished data).

Conservation International's Rapid Assessment Program (RAP)

RAP is an innovative biological inventory program designed to use scientific information to catalyze conservation action. RAP methods are designed to rapidly assess the biodiversity of highly diverse areas and to train local scientists in biodiversity survey techniques. Since 1990, RAP's teams of expert and host-country scientists have conducted over 65 terrestrial, freshwater aquatic (AquaRAP), and marine biodiversity surveys and have contributed to building local scientific capacity for scientists in 26 countries. Biological information from previous RAP surveys has resulted in the protection of millions of hectares of tropical forest, including the declaration of protected areas in Bolivia, Peru, Ecuador, and Brazil and the identification of biodiversity priorities in numerous countries.

Criteria generally considered during RAP surveys to identify priority areas for conservation across taxonomic groups include: species richness, species endemism, rare and/or threatened species, and habitat condition. Measurements of species richness can be used to compare the number of species between areas within a given region. Measurements of species endemism indicate the number of species endemic to some defined area and give an indication of both the uniqueness of the area and the species that will be threatened by alteration of that area's habitat (or conversely, the species that may be conserved through protected areas). Assessment of rare and/or threatened species (IUCN 2008) that are known or suspected to occur within a given area provides an indicator of the importance of the area for the conservation of global biodiversity. The confirmed presence or absence of such species also aids assessment of their conservation status. Many of the threatened species on IUCN's Red List carry increased legal protection thus giving greater importance and weight to conservation decisions. Describing the number of specific habitat types or subhabitats within an area identifies sparse or poorly known habitats within a region that contribute to habitat variety and therefore to species diversity.

RAP SURVEY AREAS

Site 1: Acarai Mountains, N 01° 23' 12.5" W 058° 56' 46.0"; elevation: 251 m a.s.l.
6 – 17 October 2006

The first site was situated at the base of the Acarai Mountains, located at an elevation of approximately 270 m, and characterized by sandy, oligotrophic (low nutrient levels) soils, with lowland evergreen, deciduous forests. The DBH of most trees was small (less than 50 cm), although widely scattered emergents with very large DBH (more than 100 cm) were also present at the site. It appeared that the forest here did not inundate seasonally nor did it inundate every year. Consequently, the leaf litter layer in the forest was rich in soil mesofauna, including nests of several species of fungus-growing ants. In addition to deciduous vegetation this site also included a small patch of native bamboo forest (*Guadua* sp.). A satellite camp was also established near the top of the Acarai Mountains in terra firme, or upland forest, at an elevation of approximately 500 m (see New Romeo Camp on Map). From here, the peak of the Acarais, approximately 1,100 m elevation, could be accessed but only ants and passalid beetles were collected there. The higher species richness documented at Site 1 for these groups could be a result of this elevational gradient and the greater sampling effort.

Site 2: Kamoa River, N 01° 31' 51.8" W 058° 49' 42.4"; elevation: 240 m a.s.l.
18 – 27 October 2006

The second main camp was established on the north bank of the Kamoa River at an elevation of 250 m. The lowland forest at the site was annually inundated with areas of palm swamp and extremely oligotrophic (nutrient poor), clay soils. Inland from the river, the transition from inundated forest and palm swamp to terra firme moist forest (~300 m) was clearly demarcated. The DBH of most trees at the site was small, and the leaf litter layer was poorly developed.

Additional Focal Areas: Water Quality and Fish Surveys

The RAP fish and water quality teams sampled areas near main channels of the Essequibo, Sipu and Kamoa rivers and their smaller tributaries, as well as primary waterways near the Masakenari village. Fishes and aquatic macroinvertebrates were surveyed at 18 sampling stations within five focal areas: 1) Focal Area 1 – Sipu River; 2) Focal Area 2 – Acarai Mountains; 3) Focal Area 3 – Kamoa River; 4) Focal Area 4 – Wanakoko Lake/Essequibo River; and 5) Focal Area 5 – Essequibo River at Akuthophono and Masakenari Village.

OVERALL RAP RESULTS

The data collected during the RAP survey indicate that the forests of the Konashen COCA are in very good condition and support rich biodiversity. Water quality was high, with no evidence of pollution. Typical of the forests of the Guayana Shield, abundance levels of species of most groups were low, but species diversity was remarkably high. These forests also exhibit low species endemicity, yet the potential for finding new taxa here is high due to the lack of prior scientific exploration.

During the survey the RAP team recorded the presence of 319 species of birds, including a species new to Guyana – a bamboo specialist, the Large-Headed Flatbill *(Ramphotrigon megacephalum)*. Remote camera trapping, combined with spoor tracking and interviews with Wai-Wai hunters recorded the presence of over 20 species of large mammals, including 5 species of monkeys, tapirs, giant ant eaters, and jaguars. Sixty species of amphibians and reptiles were recorded, including threatened species such as *Dendrobates azureus* (VU) and *Geochelone denticulata* (VU). Both the observed diversity and abundance of amphibians was low due to the dry season conditions during the survey, and the actual amphibian species diversity at the visited sites is probably itself greater than 70 species. Fish diversity was also affected by low water levels, but nonetheless over 100 species of fish were recorded, including 3 species of catfish likely new to science. Ant diversity was very high, with estimated numbers of 200+ species in the leaf litter alone. At least one species is likely new to science (a leaf cutting *Trachymyrmex*) and one genus *(Mycetarotes)* was recorded for the first time outside of the Amazon Basin. Katydid diversity was also high, with 72 species recorded. Of these at least seven are likely new to science and at least 57 are new records for Guyana. The results of dung beetle sampling are not finalized, but at least 50 species of beetles were collected and at least one species of passalid beetles is likely new to science.

Population levels of species hunted and fished by the Wai-Wai community were high, and did not exhibit symptoms of overharvesting, although species abundance of certain frequently hunted species (e. g., black caimans and wood tortoises) were significantly lower in the immediate vicinity of the Wai-Wai village of Masakenari.

RAP RESULTS BY TAXONOMIC GROUP

Katydids (Orthoptera)

Seventy-three (73) species of katydids (grasshopper relatives, order Orthoptera) were recorded, 58 of which (79%) are new records for Guyana, and at least seven are likely new to science. The RAP survey increased the known katydid fauna of Guyana by 130% for a total of 101 species, yet this number probably represents only about 30% of the actual diversity of katydids in the country. More katydid species were collected at Site 1 Acarai Mountains than at Site 2 along the Kamoa River. However, population densities were low at both sites, which is typical of a pristine, undisturbed primary growth forest. Almost all of the katydid species recorded during the RAP survey are indicative of undisturbed forest habitats.

Ants (Hymenoptera: Formicidae)

Because several years' time is necessary for sorting and preparing the estimated 25,000 specimens collected, we are not yet able to provide a detailed report on the results. A preliminary review indicates that ant diversity was very high, with estimated numbers of 200 species in the leaf litter alone (pending final identifications). At least one species is likely new to science (a leaf cutting *Trachymyrmex*) and one genus (*Mycetarotes*) was recorded for the first time outside of the Amazon Basin. Using hand collecting techniques, the ant team recorded 34 ant genera representing 9 subfamilies (of 21 subfamilies currently defined for the family Formicidae (see Appendix1)). Site 1 contained the larger number of genera (33), whereas Site 2 contained 22 genera. It is important to note that there was a greater sampling effort by the ant team at Site 1 than at Site 2. Both sites shared 22 genera out of the total of 34 collected. Site 1 contained a higher number of exclusive ant genera (11), i.e., genera not shared with Site 2, whereas Site 2 contained only one of the non-shared genera.

Beetles (Coleoptera: Scarabaeidae)

The results of the beetle sampling are not yet finalized, but at least 50 species of dung beetles were collected and at least one species of passalid beetles appears to be new to science. Initial impressions indicate that Site 1 (Acarai Mountains) supports a more diverse assemblage of species and genera than does Site 2 (Kamoa River). At Site 1, the following genera were observed: *Deltochilum, Ateuchus, Dichotomius, Ontherus, Canthon, Eurysternus, Oxysternon, Phanaeus,* and *Cryptocanthon.* At Site 2, no *Phanaeus, Deltochilum* nor *Cryptocanthon* were seen in the initial assessments of the traps although some specimens will likely be found when examined more carefully in the lab. The multiple elevations assessed at Site 1 support distinct scarabaeine faunae although a few species, such as *Oxysternon festivum,* were common at all sites, irrespective of elevation.

Water Quality

The water quality team measured and collected water samples from the Sipu, Essequibo and Kamoa rivers and associated tributaries, creeks and the Masakenari village well and taps. The rivers and creeks are used for domestic purposes and only in Masakenari is drinking water obtained from a well. The average water temperature was 25 °C and ranged from 23 °C to 25.9 °C. The pH values of the majority of creeks, rivers and isolated pools were similar to river water values observed in the Amazon basin and ranged from 4.74 to 6.4. These were lower than the secondary (non-enforceable) drinking water standards of the WHO or USEPA while those of the village well were close to the minimum requirement of 6.5 pH units. Alkalinity of samples was between 7.5 and 27.5 mg/L as $CaCO_3$, indicating low buffer capacity waters. For all of the samples, arsenic and aluminum concentrations were within USEPA primary drinking water standards. Basic water quality data and observations show that the main rivers and creeks of the Konashen COCA are free of human or industrial pollution.

Fishes

A total of 113 species of fishes were identified, representing six orders and 27 families. The order Characiformes (tetras, piranhas, etc.) with 61 species (51.7%) was the most diverse, followed by Siluriformes (catfishes) with 32 species (27.1%), and Perciformes (cichlids, drums) and Gymnotiformes (electric or knife fishes) with nine species each (15.3% respectively). Family Characidae contributed the most species, with 31 species collected (27.4%) followed by Loricariidae with 13 species (11.5%).

Focal Area 5 exhibited the highest fish species richness, with 53 species (46.9%), followed by Focal Areas 1 (48 species), 3 (45 species), 4 (33 species) and 2 (32 species). According to the distribution of fish species, and based on the similarity index and physicochemical variables, Focal Areas 1 and 3 exhibited the highest similarity (0.67), and can be viewed as possessing similar ichthyological communities. The remaining Focal Areas exhibited lower values, between 0.4 and 0.26, and are therefore considered to be of moderate similarity.

Nearly half of the fish species recorded are considered important subsistence fish resources; 20% are of sport fishing interest and approximately 75% have ornamental value. Four species of fishes are likely new to science: *Hoplias* sp., *Ancistrus* sp., *Rivulus* sp., and *Bujurquina* sp.

Aquatic Macroinvertebrates

Ten species of aquatic macroinvertebrates were identified, belonging to three classes (Crustacea, Gastropoda, and Bivalvia), of which Crustacea was the most diverse with three families. Of these, Palaemonidae showed the highest richness (4 species), followed by Pseudothelphusidae (3 species), and Trichodactylidae (2 species). The classes Gastropoda (snails) and Bivalvia (mussels) were represented by one species each. The greatest species richness was found in Focal Areas 2 and 3, with five and six species of aquatic macroinvertebrates respectively, while three species were collected in each of the remaining focal areas, except for Focal Area 5 where four species were recorded.

Reptiles and Amphibians

A total of 26 species of amphibians and 34 species of reptiles were recorded. The amphibians include representatives of the orders Gymnophiona (caecilians) and Anura (toads and frogs). More than half of the recorded anurans were treefrogs (Hylidae) with 13 species (54% of all recorded species), followed by the Leptodactylidae with five species. Within reptiles, 2 species of crocodilians, 3 turtles, 14 lizards and 16 snakes were recorded. The blind snake *Typhlophis ayarzaguenai* represents the first record of this species for Guyana. The aquatic lizard *Neusticurus* cf. *rudis*, the snake *Helicops* sp., and the caecilian may also represent new records for the Guyana herpetofauna, but require additional taxonomic reviews.

The three sites explored during this survey differed in the composition of the reptile and amphibian fauna. The surveyed region appears intact and in pristine condition, particularly the Acarai Mountains and the flooded forests adjacent to the main channels of the Kamoa and Sipu rivers. The area of the Essequibo River closest to Masakenari and Akuthopono villages showed a lower abundance of medium-to-large bodied reptiles, turtles and caimans, which are a part of the Wai-Wai diet, but populations of other reptiles and amphibians seemed to be in good condition.

Birds

Bird species richness was high at both RAP sites; a combined total of 319 species was tallied over the study period. The avifauna was typical of Guianan lowland forest, including 32 species endemic to the Guayana Shield. There was a high degree of habitat heterogeneity within each site, thus the avian diversity was higher than expected for the size of the area surveyed. It is probable that at least 400 bird species, or more than half of the number known to occur in Guyana, may be found in the Konashen COCA.

The survey recorded Large-headed Flatbill (*Ramphotrigon megacephalum*), a new record for Guyana and a range extension of approximately 900 km. Populations of parrots, guans, and curassows, all of which are important to the Wai-Wai inhabitants of the region and are of global conservation concern, seemed healthy. Fourteen species of parrots were observed, including Scarlet Macaw (*Ara macao*), a CITES Appendix 1 species and Blue-cheeked Parrot (*Amazona dufresniana*), considered Near Threatened (IUCN 2006). Some of the larger parrot species are hunted by local people, but the effects of this hunting appear to be negligible. Spix's Guan (*Penelope jacquacu*) and Black Curassow (*Crax alector*) were common at both survey sites, suggesting that their regional populations are not threatened by current levels of hunting pressure from the local community.

Large Mammals

Twenty-one species of large mammals were recorded during the RAP survey, with a total of 42 large mammal species expected for the area. Five species of conservation concern were recorded: the Brown-bearded saki monkey (*Chiropotes satanas*) and the Giant otter (*Pteronura brasiliensis*) listed as Endangered, and the Giant armadillo (*Priodontes maximus*), Bush dog (*Speothos venaticus*), and Brazilian tapir (*Tapirus terrestris*) considered Vulnerable (IUCN 2006). Both RAP study sites are utilized as hunting areas for two weeks per year by the local people, but otherwise appear to be pristine, undisturbed tropical rain forest. The RAP results suggest that the sites sampled contain the full complement of the large mammal species characteristic of the Guayana Shield. Because this region has a very low human population density (0.032 humans/km^2) the forests of the Konashen Indigenous District are likely to contain an intact faunal assemblage of large mammals.

CONSERVATION CONCLUSIONS AND RECOMMENDATIONS

The sites visited by the RAP team in the Konashen COCA belong to some of the most pristine and least populated areas in South America. Because the human density is low and pressure on natural resources is carefully managed by Wai-Wai community leaders, the flora and fauna of the Konashen COCA is currently intact and secure. The remoteness of the Konashen COCA has no doubt served to protect it and presently most of the area is under very little threat and anthropogenic disturbances are negligible. However, the Wai-Wai and their partners must be vigilant in protecting and managing the forests and their biodiversity to avoid species declines and to maintain the pristine condition of the forests.

I. Address Potential Threats: Prevent Illegal Logging and Mining
While there are no known factors immediately threatening the forests of the Konashen COCA, the development of roads in the neighboring regions of Brazil could result in encroachment through illegal logging or gold mining. Infrastructure development in Lethem, in the form of the Linden-Lethem road and the Takatu Bridge that will connect Brazil and Guyana, make the establishment of anti-logging/anti-mining guidelines even more timely and urgent.

1. Establish guidelines that inhibit mining and logging activities in the Konashen COCA and the wider Konashen Indigenous District.

Since illegal mining appears to pose the greatest potential threat, the Wai-Wai community should adopt a set of guidelines that exclude illegal mining and logging in the COCA. As this threat to biodiversity is likely to come from external sources far beyond the Wai-Wai community, it is essential that the Wai-Wai community leadership continues to manage their resources in a sustainable manner and prevents outsiders from jeopardizing the ecological integrity of the area.

2. Monitor the Konashen COCA to detect encroachment by illegal gold miners and loggers.

The ongoing construction of a highway across northern Brazil will likely exacerbate the current problems associated with illegal miners in the interior of the Guianas. If necessary, the Wai-Wai should enlist the help of the Government of Guyana and other partners to implement a program of patrols to discourage illegal miners and loggers from entering their territory. Frequent patrols of the borders of the COCA may be necessary to detect encroachment.

II. Sustainably Manage Natural Resources

The Wai-Wai community of Masakenari in the Konashen COCA is Guyana's most remote village and residents have minimal external contacts. As such, most of the raw materials for their food, housing, craft and medication come from the area. In addition, the major form of economic activity for the Wai-Wai is international wildlife trade. As good custodians of their environment, the Wai-Wai residents have recognized that their needs for resources are increasing and if not well managed there may be serious negative impacts in the future. A management plan for the Konashen COCA is in development. Recommendations from the RAP survey include the following related to 1) Sustainable Harvesting, 2) Monitoring, 3) Species Protection, and 4) Capacity Building. The Wai-Wai currently hunt and harvest a wide variety of forest animals, including fishes, large reptiles, birds, and mammals. Some of the species often hunted or harvested by the Wai-Wai that should be the target of the activities described below are listed in Table 1.

Sustainable Harvesting

1. Conduct further studies of key species most utilized by the Wai-Wai.

The species most hunted and harvested by the Wai-Wai should be intensively studied to determine their current population sizes and their distributions within the COCA, and to evaluate the extent to which they can be sustainably harvested in each part of the COCA. This can be done by developing and continuing collaboration between the Wai-Wai community members and scientists who can advise on research techniques, data analysis and harvesting analyses. The Wai-Wai rangers who have received training from CI-Guyana and Iwokrama should continue to be trained in further techniques so that they can conduct the research and monitoring programs.

2. A sustainable management plan should be designed and implemented, using the data from the RAP survey and additional studies recommended above.

A management and sustainable-use plan should be developed for each species that is heavily hunted or harvested by the Wai-Wai. Research by the Wai-Wai for development of the Management Plan for the Konashen COCA (in prep) includes maps of the Wai-Wai fishing and hunting grounds within the COCA. The Wai-Wai currently harvest about 20

fish species from 13 fishing grounds within the COCA using harpoons (bow and arrow) and seines. The effectiveness and impacts of current fishing practices should be evaluated and managed. Mammals are usually hunted within 15 km of Masakenari using traps, shotguns and arrows. Plant and other resources, including stones, Brazil nuts, fruits and wood for building are also harvested along the main rivers.

3. Develop and implement a rotation system to distribute the effects of subsistence hunting over as large an area as possible.

Hunting should be done judiciously by distributing hunting activity over as large an area as possible, such that the majority of the Konashen area is not used for hunting at any given time. This simple system would ensure that local populations have time to recover following brief periods of intense hunting. Cracids (guans and curassows) are arguably

Table 1. Species often hunted and harvested by the Wai-Wai in the Konashen COCA.

Fishes
Haimara (*Hoplias macrophthalmus*)
Tiger fish (*Pseudoplatystoma fasciatum*)
Kururú (*Curimata cyprinoides*)
Pakuchí or Catabact pacú (*Myleus rhomboidalis*)
Reptiles
Dwarf Caiman (*Paleosuchus trigonatus*)
Black Caiman (*Melanosuchus niger*)
Tortoises: *Rhinoclemmys punctularia*
Chelonoidis carbonaria
Chelonoidis denticulata
Birds
Scarlet Macaw (*Ara macao*) - for international trade
Red-and-green Macaw (*Ara chloropterus*)
Blue-and-yellow Macaw (*Ara ararauna*)
Blue-cheeked Parrot (*Amazona dufresniana*)
Orange-winged Parrot (*Amazona amazonica*)
Red-fan Parrot (*Deroptyus accipitrinus*)
Spix's Guan (*Penelope jacquacu*)
Black Curassow (*Crax alector*)
Gray-winged Trumpeter (*Psophia crepitans*)
Mammals
Monkeys (7 species)
Golden-handed tamarin (*Saguinus midas*)
Brazilian tapir (*Tapirus terrestris*)
Deer (*Mazama* spp. and *Odocoileus virginianus*)
Paca (*Agouti paca*)
Agouti (*Dasyprocta agouti*)
Collared and White-lipped peccaries (*Tayassu* spp.)

the most important birds in the diet of the Wai-Wai. They have low reproductive rates and tend to disappear when subjected to heavy hunting pressure. The cracid populations in Konashen are currently healthy, and it is likely that local population depletion (due to hunting) is a temporary phenomenon in most cases.

4. If necessary, establish hunting/collecting quotas using the data from the aforementioned population monitoring studies.

Assistance from expert scientists and natural resource managers would be helpful in determining and developing limits and quotas, if they become necessary, on the number of animals hunted/harvested for the COCA.

Monitoring

1. Develop and implement a plan to monitor and assess the species populations that are the Wai-Wai community's most valuable food, hunting and trade resources.

Though the current populations appear to be secure, the Wai-Wai community should implement and manage a long-term monitoring program to detect any changes in the occurrence or abundance of the species listed above, especially those that are listed by IUCN as threatened or by CITES as of concern in international trade.

The Wai-Wai rangers should continue their camera-trapping, fish and bird monitoring programs and analyze the data to detect trends and predict future scenarios. The data should be published and also made accessible to those involved in conservation and development in Guyana. A database of the biological information for the Konashen COCA should be developed and maintained within Guyana.

2. Establish a water quality monitoring program of the major rivers of the COCA.

Such a program should be conducted on a quarterly basis at selected sampling sites along the Essequibo River and in the village. This quarterly monitoring should provide two samples each, during the dry and wet seasons. Water quality monitoring should be extended to include microbial analysis in the more populated area. Sites used to collect aquatic species as well as any new sites identified by the Wai-Wai community should also be monitored for water quality on at least an annual basis.

Well water should be monitored on a consistent basis as the well sits downstream of the village garbage holes and latrines, which are unlined. In Akuthopono, water is collected from the river, and garbage is dumped in a hole used by the village as a well until the year 2000. This activity could potentially affect the groundwater quality and plans should be made to provide clean drinking water at the site if it is to be developed as an income-generating ecotourism visitor center.

Species protection

1. Continue to avoid trapping parrots for the pet trade, and deny trappers entry to the Konashen COCA.

The Guianas contribute a substantial number of parrots to the international pet trade, and trappers often travel great distances to harvest the most valuable species. This has led to dramatic declines in the populations of some species in Guyana in accessible areas closer to the coastal plain than the Konashen COCA. The remoteness of the Konashen COCA has no doubt served to protect it from such exploitation. Parrots and large game birds, though not currently threatened at a regional level, are of global conservation concern. Care should be taken to forestall local declines in their populations. Monitoring of parrots and large game birds is <u>not</u> recommended at the present time, since these species are not amenable to standardized survey methods.

2. Avoid hunting and harvesting threatened species.

Species that are considered threatened on the IUCN Red List of Threatened Species (2008) or on the CITES (2008) list of species at risk due to international trade should not be hunted or harvested. Threatened species documented during the RAP in the Konashen COCA are listed in the Report at a Glance of this report. Additional species of conservation concern, such as the Harpy Eagle (*Harpia harpyia*), were not documented during the RAP survey but are likely present in the Konashen COCA and should also be protected.

Scientific Capacity Building

1. Further develop and continue the formal training of Wai-Wai rangers and parabiologists in the study, conservation and management of aquatic and terrestrial resources.

Continued and expanded training of the local Wai-Wai community in the study, management, conservation and valuation of their biological resources will be valuable in both species population monitoring and ecotourism projects. Development of a scientific research station in the Konashen COCA would greatly enhance the conservation and research potential of the area and serve as a local education facility for the rangers and parabiologists. It is a very attractive region for researchers and scientists and would provide another source of sustainable revenue for the area. Collaborations between the Wai-Wai rangers and expert scientists should be sought and developed to carry out in-depth studies of threatened species and species harvested by the Wai-Wai.

III. Promote Sustainable Ecotourism

1. Implement a basic eco-tourism infrastructure that supports research and education-based activities that will further enhance conservation efforts and biological knowledge of the Konashen Indigenous District.

The pristine ecological condition of the forests of the Konashen Indigenous District supports great potential for research-based opportunities and education-based ecotourism. Such activities could generate revenue for the area's inhabitants while simultaneously encouraging biodiversity protection and scientific research in the area. The largely undisturbed habitat and dazzling array of charismatic species such as the giant otter, caimans, tapir, large cats, the haimara and the outstanding bird diversity are major draws for eco-tourists around the world. With continued formal training for the rangers and parabiologists, ecotourism of the area could easily be promoted in the Konashen COCA.

Plans should be made to provide clean drinking water at the site if it is to be developed as an income-generating ecotourism visitor center. Plans should also be made to quantify the water resources in the area, especially since the area experiences high levels of rainfall and is inundated for large periods of the year. This information will also assist with safer plans for water and sanitation in Masakenari and Akuthupono. Long-term monitoring of the environmental impact of ecotourism should be put in place to ensure the most sustainable practices are promoted and can be used to "certify" the area as an ideal destination for "sustainable ecotourism."

In addition to the larger, well-known species of the Konashen COCA such as the sloth and primates, many invertebrates like the spectacular and rather common Peacock katydid (*Pterochroza ocellata*) or *Morpho* butterflies, have the potential to attract ecotourism. It is therefore important to continue training Wai-Wai parabiologists in recognizing some of the more iconic and "charismatic" invertebrates, which are becoming popular targets of the ecotourism industries in other parts of the world. Many of the amphibians and reptiles recorded during the RAP survey are also of great eco-tourism potential and/or are important in the pet trade.

IV. Conduct Further Research

1. Conduct further, more extensive sampling during the rainy and dry seasons, paying particular attention to the aquatic biodiversity. Additional research needs have been identified for:

Fishes: The lower section of the Essequibo River, from Masakenari to the Amaci Falls, is of great diversity and use to the Wai-Wai, and remains to be sampled. For this reason, it is fundamental to conduct a second sampling expedition in the low water season (November-December) on the Wai-Wai fishing grounds which include, but are not limited to Amaci Falls, Kanaperu, Mekereku and Wanakoko. This would result in a more comprehensive and accurate species list, particularly in regard to the smaller-sized species.

Among the fish species identified during the RAP survey, many species have high potential for aquarium and ornamental trade. However, to develop a plan that is sustainable and effective would require additional information on the present species' distribution and abundance. Taking this into account, it is recommended to complete an inventory of the fish species, and subsequently continue biological, ecological and market studies of these species.

Reptiles and Amphibians: The results of this survey are preliminary, and we suspect that a much greater diversity of amphibians and reptiles is to be found here. For this reason, we recommend more extensive sampling of the entire region, including sampling during both rainy and dry seasons. Also, particular attention should be given to the Acarai Mountains where we expect a high species richness and a possible center of endemism for amphibians and small reptiles. Specific studies of the use of large reptiles (e.g., black caimans and tortoises) by the Wai-Wai are also needed to develop sustainable harvesting plans.

Water Quality and Resources Assessment: Plans should be made to quantify the water resources in the area, especially since the area experiences high levels of rainfall and is inundated for large periods of the year. This information will also assist with safer plans for water and sanitation in Masakenari and Akuthopono.

Invertebrates: Based on the results of the katydid survey, we strongly recommend additional entomological surveys of the Konashen District, which are bound to yield many species of insects and other invertebrates that are new to science.

REFERENCES

Braun, M.J., D.W. Finch, M.B. Robbins and B.K. Schmidt. 2000. A Field Checklist of the Birds of Guyana. Smithsonian Institution, Washington, DC.

CITES. 2008. Convention on International Trade in Endangered Species. www.cites.org.

Engstrom, M. and B. Lim. 2008. Checklist of the Mammals of Guyana. Online checklist <http://www.mnh.si.edu/biodiversity/bdg/guymammals.html>, accessed May 14, 2008.

Funk, V., T. Hollowell, P. Berry, C. Kelloff, and S.N. Alexander. 2007. Checklist of the Plants of the Guiana Shield (Venezuela: Amazonas, Bolivar, Delta Amacuro; Guyana, Surinam, French Guiana). Contributions from the United States National Herbarium, volume 55.

Hammond, D. 2005. Ancient Land in a Modern World. *In:* Tropical Forests of the Guiana Shield: ancient forests of the modern world (Ed. D. Hammond). CABI International, Oxford, UK. pp 1-14.

Hollowell, T., and R. P. Reynolds (eds.). 2005. Checklist of the Terrestrial Vertebrates of the Guiana Shield. Bulletin of the Biological Society of Washington, no. 13.

Hollowell, T, P. Berry, V. Funk and C. Kelloff. 2001. Preliminary Checklist of the Plants of the Guiana Shield (Venezuela: Amazonas, Bolívar, Delta Amacuro; Guyana; Surinam; French Guiana). Volume 1: Acanthaceae – Lythraceae. http://www.mnh.si.edu/biodiversity/bdg/guishld/index.html.

Huber, O. and M.N. Foster. 2003. Conservation Priorities for the Guayana Shield: 2002 Consensus. Conservation International. Washington, DC.

IUCN. 2008. IUCN Red List of Threatened Species. www.redlist.org.

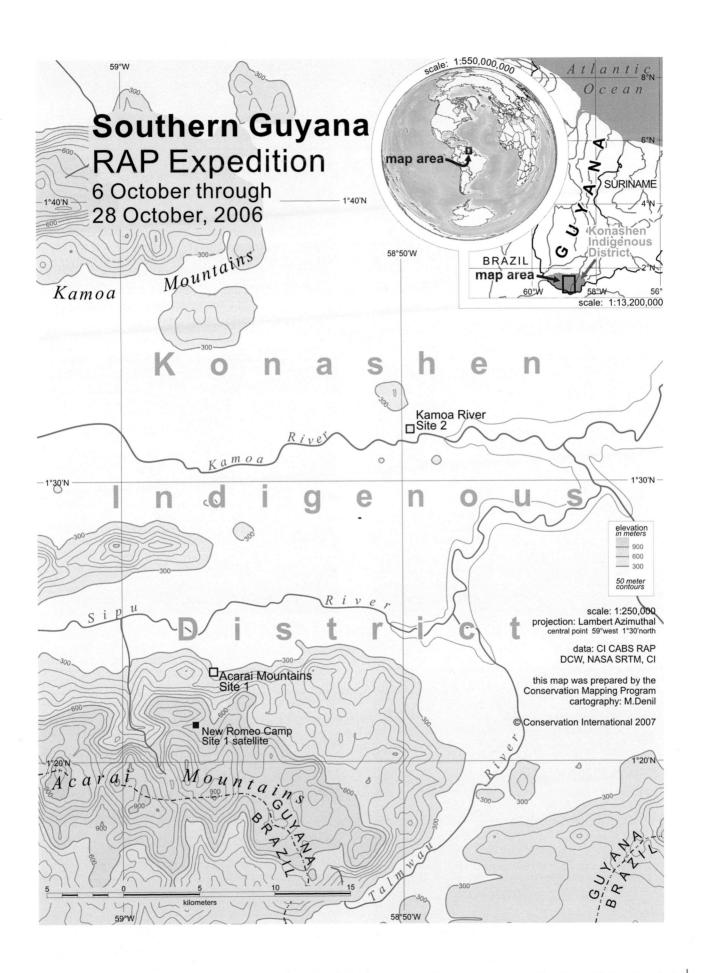

Southern Guyana
RAP Expedition
6 October through
28 October, 2006

scale: 1:550,000,000

Atlantic Ocean

map area

GUYANA
SURINAME
Konashen Indigenous District
BRAZIL
map area

scale: 1:13,200,000

Kamoa

Mountains

K o n a s h e n

Kamoa River
Site 2

River

Kamoa

I n d i g e n o u s

River

Sipu

D i s t r i c t

Acarai Mountains
Site 1

New Romeo Camp
Site 1 satellite

Acarai *Mountains*

GUYANA
BRAZIL

Talmwau

River

GUYANA
BRAZIL

elevation
in meters
900
600
300

50 meter contours

scale: 1:250,000
projection: Lambert Azimuthal
central point 59°west 1°30'north

data: CI CABS RAP
DCW, NASA SRTM, CI

this map was prepared by the
Conservation Mapping Program
cartography: M.Denil

© Conservation International 2007

kilometers

Large-headed Ant (*Daceton armigerum*) showing large mandibles

Freshwater Crab (*Fredius* sp.) can sometimes be found outside of water, foraging on the forest floor

Blue Poison Dart Frog (*Dendrobates tinctorum*)

Mouse possum (*Marmosa* sp.)

Forest floor of a typical, annually inundated forest along the Kamoa River

Emerald Tree Boa (*Corallus caninus*) coiled in a tree

Ichthyologist Carlos Lasso with a freshly caught catfish

Surinam Toad (*Pipa pipa*) mimicking leaf litter at the bottom of a forest pond

Tukeit Hill Frog *(Allophryne ruthveni)*

Bess Beetle (*Passalus* sp.)

Peacock Katydid (*Pterochroza ocellata*) amid leaf litter

Lobster Katydid (*Panoploscelis specularis*), one of the largest katydids of the Neotropics

A pristine stream in the Acarai Mountains, the first site of the RAP survey

Worm Lizard (*Amphisbaena vanzolinii*) is legless and without functional eyes

Smooth-fronted Caiman (*Paleosuchus trigonatus*) young, this species prefers more turbid waters than other species of the genus

Giant Bird-eating Spider (*Theraphosa blondi*), the largest spider in the world

Entomologist Christopher Marshall and assistants collecting symbiotic moths from the fur of Pale-throated Three-toed Sloth (*Bradypus tridactylus*)

Chapter 1

Katydids of selected sites in the Konashen Community Owned Conservation Area (COCA), Southern Guyana

Piotr Naskrecki

SUMMARY

The survey of katydids (Orthoptera: Tettigoniidae) of the Konashen COCA of southern Guyana revealed a high species diversity of these insects, resulting in a 130% increase of the katydid fauna of this country. Seventy-three species were recorded, 58 of which (79%) were new to Guyana, and at least seven were new to science. Combined with 44 species previously recorded, the known katydid fauna of Guyana now includes 101 species, yet this number probably represents only about 30% of the actual diversity of these insects. Virtually all species recorded during this survey are indicative of undisturbed forest habitats.

INTRODUCTION

Despite the recent increase in the faunistic and taxonomic work on katydids (Orthoptera: Tettigonioidea) of the northern Neotropics, forests and savannas of the Guayana Shield remain some of the least explored, and potentially most interesting regions of South America. Collectively, over 190 species of the Tettigoniidae have been recorded from the countries comprising the Guayana Shield (Venezuela, Guyana, Suriname, and French Guiana), but this number clearly represents only a fraction of the actual species diversity in this area. Most of the known species were described in the monumental works by Brunner von Wattenwyl (1878, 1895), Redtenbacher (1891), and Beier (1960, 1962). More recently Nickle (1984), Emsley and Nickle (2001), Kevan (1989), and Naskrecki (1997) described additional species from the region. Overall, forty-four species of katydids have been recorded from Guyana, but it is likely, based in part on the result of this survey, that at least 250-300 species occur there.

Katydids have long been recognized as organisms with a significant potential for their use in conservation practices. Many katydid species exhibit strong microhabitat fidelity, low dispersal abilities (Rentz 1993), and high sensitivity to habitat fragmentation (Kindvall and Ahlen 1992) thus making them good indicators of habitat disturbance. These insects also play a major role in many terrestrial ecosystems as herbivores and predators (Rentz 1996). It has been demonstrated that in the neotropical forests katydids are themselves a principal prey item for several groups of invertebrates and vertebrates, including birds, bats (Belwood 1990), and primates (Nickle and Heymann 1996). While no neotropical species of katydids has been classified as threatened (primarily because of the paucity of data on virtually all species known from this region), there are already documented cases of species of nearctic katydids being threatened or even extinct (Rentz 1977).

The following report presents the results of a Rapid Assessment survey (RAP) of katydids conducted between October 7-27, 2006 at selected sites within the Konashen Indigenous District of southern Guyana. All collecting sites of the survey were located within the boundaries of the Community Owned Conservation Area (COCA), a protected area belonging to the Wai-Wai community. The katydid fauna of this area, similar to that of most of the country, has never been surveyed, and most species found during the RAP survey represent new records for Guyana.

METHODS AND STUDY SITES

During the survey three collecting methods were employed for collecting katydids: (1) collecting at mercury vapor (MV) and ultraviolet (UV) lights at night, (2) visual search at night, and (3) net sweeping of the understory vegetation during the day and at night. An ultrasound detector (Pettersson D 200) was also used to locate species that produce calls in the ultrasonic range, undetectable to the human ear. MV trapping was done for 10 nights (5 nights at each main site), for at least 3 continuous hours between 19:30 and midnight. UV trapping was done sporadically, and only at sites (Sites 3 and 4) where using a generator to power the MV lamp was not feasible (the UV lamp could be powered by a relatively small, rechargeable 12 V battery.) Visual searches were done every night, usually between 20:00 and 02:00, when the activity of virtually all katydid species was the highest. Vegetation sweeping was done only during the day, and only in places where the density of the vegetation permitted such an activity (e.g., grasslands near the Wai-Wai villages, and along river banks). Sweeping was standardized by performing five consecutive sweeps in a series before the content of the net was inspected.

Representatives of all encountered species were collected and voucher specimens were preserved in 95% alcohol, or as dry, pinned specimens. Upon completion of their identification, most of the specimens will be returned to the collection of the University of Guyana, Georgetown, while a small number of voucher specimens will be deposited in the collections of the Museum of Comparative Zoology, Harvard University, and the Academy of Natural Sciences of Philadelphia (the last will also become the official repository of the holotypes of possible new species encountered during the present survey upon their formal description).

In addition to physical collection of specimens, stridulation of acoustic species was recorded using the Sony MZ-NHF 800 digital recorder and a Sennheiser shotgun microphone. These recordings are essential to establish the identity of potentially cryptic species, in which morphological characters alone are not sufficient for species identification. Virtually all encountered species were photographed, and these images will be available online in the database of the world's katydids (Eades et al. 2007).

The majority of specimens were collected at the following two sites:

1. Site 1: Acarai Mountains – Foothills of the Acarai Mountains, lowland, terra firme forest along Acarai creek, N01°23'12.2", W058°56'45.7", elevation ~270 m; collecting was done between October 6th and 17th, 2006. This site was characterized by sandy, oligotrophic soils, with lowland evergreen, deciduous forests. The DBH of most trees was small (less than 50 cm), although widely scattered emergents with very large DBH (more than 100 cm) were also present at the site. It appeared that the forest here did not inundate seasonally nor did it inundate every year. Consequently, the leaf litter layer in the forest was rich in soil mesofauna, including a number of species of katydid associated with leaf litter (e.g., *Uchuca* spp.). In addition to deciduous vegetation, this site also included a small patch of native bamboo forest (*Guadua* sp.).

2. Site 2: Kamoa River – A lowland, seasonally inundated forest along the Kamoa River, N01°31'52", W058°49'41.9", elevation ~250 m; collecting was done between October 18th and 27th. The forest at the site was annually inundated with areas of palm swamp and extremely oligotrophic clay soils. Inland from the river, the transition from inundated forest and palm swamp to terra firme moist forest (~300 m) was clearly demarcated. The DBH of most trees at the site was small, and the leaf litter layer was poorly developed.

Additionally, opportunistic collecting was done at two sites near the main RAP campsites:

3. Akuthopono (nr. Gunns landing strip) – Mostly savanna, with small patches of riparian forest, N01°39'04.2", W058°37'42.9", elevation ~230 m; collecting was done on October 3rd and 27th.

4. Sipu River campsite – A landing site on the Sipu River, inundated lowland forest, N01°25'06", W058°57'12.6", elevation ~250 m; collecting was done on October 3rd.

RESULTS

The katydid fauna encountered during the Konashen RAP survey turned out to be exceptionally rich, both in terms of the numbers of species recorded (73), and species new to Guyana (58, or 79% of recorded species; Table 1.1). At least seven species have been confirmed to be new to science, but additional, yet unidentified species may prove to be new as well (although still unidentified, they are not conspecific with species previously recorded from Guyana). Combined with the 44 species already recorded from Guyana, the number of katydid species known from this country is now 101, nearly a 130% increase.

Virtually all species recorded during this survey are forest species, known only from undisturbed forests of the Guyana Shield. Only one species, *Neoconocephalus purpurascens,* is also associated with open, grassy habitats.

Most members of the Phaneropterinae, a subfamily that includes a large proportion of volant, canopy species, were collected with the use of the MV or UV light traps, whereas virtually all Pseudopyllinae, many of which are non-volant, were collected in the understory during visual night collecting. The Listroscelidinae were collected mostly by vegetation sweeping and at lights. The Conocephalinae include both volant and non-volant species, found in the canopy, the understory, or even in the leaf litter, and consequently were collected using all three main collecting methods.

Table 1.1. A checklist of katydid species collected in the Konashen COCA, Southern Guyana.

Species	Site 1 Acarai Mtns	Site 2 Kamoa River	Site 3 Akuthopono	Site 4 Sipu River	Likely new to science	New record for Guyana
Conocephalinae						x
Copiphora gracilis						x
Daedalellus apterus		x	x			x
Eschatoceras sp. n. 1	x	x			x	x
Gryporhynchium acutipennis	x					x
Lamniceps sp. 1		x			x	x
Neoconocephalus purpurascens	x	x	x			
Paralobaspis gorgon	x	x				x
Subria.grandis		x				x
Uchuca similis	x	x				x
Vestria diademata	x	x				x
Listroscelidinae						
Listroscelis armata	x	x				x
Phlugiola sp. 1	x	x			x	x
Phlugis cf, *bimaculatoides*	x	x	x	x		x
Phaneropterinae						
Anaulacomera sp. 1	x					x
Anaulacomera sp. 2	x					x
Anaulacomera sp. 3	x					x
Anaulacomera sp. 4	x					x
Anaulacomera sp. 5	x	x				x
Ceraia sp. 1	x	x				x
Ceraia sp. 2	x	x				x
Ceraia sp. 3		x				x
Euceraia subaquila	x					x
Euceraia sp. 1	x					x
Ceraiaella sp. 1	x					x
Euceraia rufovariegata	x					x
Gen. 1 sp. 1	x				x	x
Gen. 2 sp. 1	x					x
Gen. 3 sp. 2	x					x
Gen. 4 sp. 1	x	x				x
Hyperphrona bidentata	x					x
Ischyra sp. 1	x	x				x
Itarissa sp. 1	x	x				x
Paraphidnia verrucosa	x					x
Phylloptera sp. 1	x					x
Pycnopalpa sp. 1	x					x
Steirodon dentiferoides	x					x

Species	Site 1 Acarai Mtns	Site 2 Kamoa River	Site 3 Akuthopono	Site 4 Sipu River	Likely new to science	New record for Guyana
Steirodon maroniensis	x					x
Steirodon sp. 3		x				x
Syntechna sp. 1	x					x
Vellea cruenta		x				x
Viadana sp. 1	x					x
Pseudophyllinae						
Acanthodis longicauda	x					
Bliastes contortipes	x	x				
Chondrosternum triste	x	x				
Chondrosternum sp. 1	x	x				x
Cycloptera speculata	x	x				
cf. *Leptotettix* sp. 1	x	x				x
Eubliastes adustus	x					x
Eumecopterus nigrovittatus	x	x				
Gen. 5 sp. 1	x					x
Gnathoclita vorax	x					
Leptotettix sp. 1	x	x				x
Leptotettix spinoselaminatus	x	x				
Leurophyllum consanguineum	x	x				
Leurophyllum sp. 1	x					
Leurophyllum sp. 2	x					
Panoploscelis scudderi	x					
Parapleminia sp. 1	x					x
Pezochiton sp. 1	x	x				x
Platychiton brunneus	x	x				
Platyphyllum sp. 1	x				x	x
Platyphyllum sp. 2	x	x			x	x
Platyphyllum sp. 3		x			x	x
Pleminia sp. 1	x					x
Pleminia sp. 2		x				x
Pleminia sp. 3		x				x
Pterochroza ocellata	x	x		x		
Rhinischia surinama	x					
Roxelana crassicornis	x					x
Scopioricus latifolius .		x				x
Diacanthodis granosa .	x					
Typophyllum rufifolium	x	x		x		x
Typophyllum flavifolium	x					x
Totals						
73	62	39	3	3	7	57

Of the two main collecting sites, Site 1 (Acarai Mtns.) yielded a higher species count (62) than Site 2 (Kamoa River) (39), but this difference should probably be attributed to less than optimal placement of the MV light trap at the second site, rather than differences in the habitat quality of vegetation (most species missing from Site 2 are volant, canopy species that can only be collected at lights). Below I discuss some of the most interesting finds of this katydid survey.

Subfamily Conocephalinae

The Conocephalinae, or the conehead katydids, include a wide range of species found in both open, grassy habitats, and high in the forest canopy. Many species are obligate graminivores (grass feeders), while others are strictly predaceous. A number of species are diurnal, or exhibit both diurnal and nocturnal patterns of activity. Ten species of this family were recorded.

Eschatoceras sp. n. 1 – Seven species of this genus are known, ranging in their distribution from Ecuador through Suriname and Brazil to Bolivia. The specimens collected at the sites within the Konashen COCA represent an additional, eighth species. They are similar to *E. bipunctatus* (Redtenbacher) but differ in the pattern of facial markings and the male genitalic structures.

Neoconocephalus purpurascens – Species of the genus *Neoconocephalus* are nearly always associated with open, grassy or marshy habitats. Most species are seed-feeders, and females lay their eggs in stems of grasses or reeds. Very few species of this large genus (120 valid species) are known to occur in forested habitats, and then only if there are openings or roads intersecting the forest. Finding *N. purpurascens* in continuous, pristine forests of the Konashen COCA adds an interesting element to the biology of this group of katydids.

Subfamily Listroscelidinae

Within the Neotropics this family is represented by over 70 species, all obligatory predators of other insects. With the exception of the genus *Phlugis*, some of which can be found in anthropogenic, grassy habitats, all species of this family appear to be associated with undisturbed, primary forests. During the present survey three species of this family were found.

Phlugiola sp. 1 – Three species of this genus are known from Peru and Suriname, all brachypterous. The individuals collected during this survey were fully winged, and represent a fourth, new species of the genus.

Subfamily Pseudophyllinae

Virtually all members of tropical Pseudophyllinae, or sylvan katydids, can be found only in forested, undisturbed habitats, and thus have a potential as indicators of habitat changes. These katydids are mostly herbivorous, although opportunistic carnivory was observed in some species (e.g.,

Panoploscelis). Many are confined to the upper layers of the forest canopy, and never come to lights, making it difficult to collect them. Fortunately, many of such species have very loud, distinctive calls, and it is possible to document their presence based on their calls alone, a technique known well to ornithologists. Thirty-two species of this family were collected during the present survey.

Gnathoclita vorax – This spectacular species is a rare example of a katydid with strong sexual dimorphism represented by a strong, allometric growth of the male mandibles. It was found at Site 1 (Acarai Mtns.) in the patches of the native bamboo (*Gaudua* sp.), where males stridulated from within bamboo stems. Until now such an association with *Gaudua* has been known only in the katydid *Leiobliastes laevis* Beier from Peru (Louton et al. 1996).

Platyphyllum spp. – Three species of this genus were recorded during the survey, all appearing to be new to science. The genus *Platyphyllum* has never been recorded from the countries of the Guayana Shield, and finding three new species in Guyana hints at a high, undiscovered diversity of this taxon.

Subfamily Phaneropterinae

Twenty-eight species of this subfamily were recorded during the survey, most of them collected at MV and UV light traps. These katydids are almost always excellent fliers, and many reside exclusively in the canopy of the forest, completing their entire reproductive cycle there. All known species of this subfamily are strictly herbivorous.

The taxonomy of Neotropical Phaneropterinae lags behind other groups of katydids, and most genera require detailed revisions before identification of most of the collected species can be completed. Nonetheless, at least one species (Gen. 1 sp. 1) has been confirmed to be new to science, and all 28 are new to Guyana, thanks to the fact that no Phaneropterinae species have ever been reported in the literature from this country.

CONSERVATION RECOMMENDATIONS

The sites visited by the RAP team in the Konashen COCA belong to some of the most pristine, least populated areas in South America. Currently, there are no known factors immediately threatening these forests, although the development of roads in the neighboring regions of Brazil may result in bringing in illegal logging or gold mining.

Based on the results of the katydid survey, I strongly recommend additional entomological surveys of the Konashen COCA, which are bound to yield many species of insects and other invertebrates that are new to science. Some of these, like the spectacular and rather common Peacock katydid (*Pterochroza ocellata*) or *Morpho* butterflies, have the potential to attract ecotourism as much as larger, better known animals. It is therefore important to continue training Wai-Wai parabiologists in recognizing some of the more iconic and "charismatic" invertebrates, which are becoming popular targets of the ecotourism industries in other parts of the world. At the same time, it would be beneficial to both the Wai-Wai community, and the scientific community in Guyana and elsewhere, to continue a biodiversity surveying and monitoring program in the Konashen COCA.

REFERENCES

Beier, M. 1960. Orthoptera Tettigoniidae (Pseudophyllinae II). *In:* Mertens, R., Hennig, W. and Wermuth, H. (eds.) Das Tierreich. 74: 396 pp; Berlin (Walter de Gruyter & Co.).

Beier, M. 1962. Orthoptera Tettigoniidae (Pseudophyllinae I). *In:* Mertens, R., Hennig, W. and Wermuth, H. (eds). Das Tierreich. 73: 468 pp.; Berlin (Walter de Gruyter & Co.).

Belwood, J.J. 1990. Anti-predator defences and ecology of neotropical forest katydids, especially the Pseudophyllinae. Pps 8–26. *In:* Bailey, W.J. and Rentz, D.C.F. (eds.) The Tettigoniidae: biology, systematics and evolution. Bathurst (Crawford House Press) & Berlin (Springer).

Brunner von Wattenwyl, C. 1878. Monographie der Phaneropteriden. 1–401, pls 1–8; Wien (Brockhaus).

Brunner von Wattenwyl, C. 1895. Monographie der Pseudophylliden. IV + 282 pp. [+ X pls issued separately]; Wien (K.K. Zoologisch–Botanische Gesellschaft).

Eades, D.C., Otte, D. and Naskrecki, P. 2007. Orthoptera Species File Online. Version 2.7. [January 2007]. <http://osf2.orthoptera.org>

Emsley, M.G. and Nickle, D.A. 2001. New species of the Neotropical genus *Daedalellus* Uvarov (Orthoptera: Tettigoniidae: Copiphorinae). Transactions of the American Entomological Society 127: 173–187.

Kevan, D.K.McE. 1989. A new genus and new species of Cocconotini (Grylloptera: Tettigonioidea: Pseudophyllidae: Cyrtophyllinae) from Venezuela and Trinidad, with other records for the tribe. Boletín de Entomología Venezolana 5: 1–17.

Kindvall, O. and Ahlen, I. 1992. Geometrical factors and metapopulation dynamics of the bush cricket, *Metrioptera bicolor* Philippi (Orthoptera: Tettigoniidae). Conservation Biology 6: 520–529.

Louton, J., J. Gelhaus and R. Bouchard. 1996. The Aquatic Macrofauna of Water-Filled Bamboo (Poaceae: Bambusoideae: *Guadua*): Internodes in a Peruvian Lowland Tropical Forest. Biotropica 28: 228-242.

Naskrecki, P. 1997. A revision of the neotropical genus *Acantheremus* Karny, 1907 (Orthoptera: Tettigoniidae: Copiphorinae). Transactions of the American Entomological Society 123: 137–161.

Nickle, D.A. 1984. Revision of the bush katydid genus *Montezumina* (Orthoptera; Tettigoniidae; Phaneropterinae). Transactions of the American Entomological Society 110: 553–622.

Nickle, D.A. and Heymann E.W. 1996. Predation on Orthoptera and related orders of insects by tamarin monkeys, *Saguinus mystax* and *S. fuscicollis* (Primates: Callitrichidae), in northeastern Peru. Journal of the Zoological Society 239: 799-819.

Redtenbacher. 1891. Monographie der Conocephaliden. Verhandlungen der Zoologisch-Botanischen Gesellschaft Wien 41(2): 315-562.

Rentz, D.C.F. 1977. A new and apparently extinct katydid from antioch sand dunes (Orthoptera: Tettigoniidae). Entomological News 88: 241–245.

Rentz, D.C.F. 1993. Orthopteroid insects in threatened habitats in Australia. Pps 125–138. *In:* Gaston, K.J., New, T.R. and Samways, M.J. (eds.). Perspectives on Insect conservation. Andover, Hampshire (Intercept Ltd).

Rentz, D.C.F. 1996. Grasshopper country. The abundant orthopteroid insects of Australia. Orthoptera; grasshoppers, katydids, crickets. Blattodea; cockroaches. Mantodea; mantids. Phasmatodea; stick insects. Sydney, University of New South Wales Press.

Chapter 2

Ants of Southern Guyana - a preliminary report

Ted R. Schultz and Jeffrey Sosa-Calvo

INTRODUCTION

Due to the combined efforts of a global community of ant systematists and ecologists, codified in Agosti et al. (2000), ants that inhabit the leaf litter have been widely used as biological indicators in biodiversity studies conducted at localities across the planet. Leaf-litter ants serve as good biodiversity indicators for conservation planning because they are: (1) ecologically dominant in most terrestrial ecosystems; (2) easily sampled in sufficiently statistical numbers in short periods of time (Agosti et al. 2000); and (3) sensitive to environmental change due to their many interdependencies with other components of the local biota (Kaspari and Majer 2000).

The ant diversity of Guyana is largely unknown. Previous studies suggest a rich ant fauna with more than 350 species (Wheeler 1916, 1918; Weber 1946; Kempf 1972; La Polla et al. 2007), but this figure likely underestimates the true number of species present in the country. It is well known that the New World tropics possess one of the richest ant faunas in the world, with nearly 3100 known species (Kempf 1972, Fernandez and Sendoya 2004).

This preliminary report summarizes ant collecting at two sites in southern Guyana, near the border of Brazil. These data will later be compared with pre-existing data from other sites across the Guayana Shield.

METHODS

Study site

The Acarai Mountains are wet, forested, low-altitude (<1500 m) uplands located in the southern part of Guyana. The Acarai Mountain range lies along the border shared between Guyana and Brazil, and is one of four mountain ranges in Guyana. Two important Guyanese rivers, the Essequibo River (the longest river in the country and the third largest river system in South America) and the Courantyne River, originate in the Acarai Mountains. The Acarai Mountains are actually one part of a larger range that extends into the Wassarai Mountains to the north and east.

Field methods

Ants were sampled at two main sites; Site 1 in the Acarai Mountains (October 6-19, 2006; Acarai Mountains including a satellite camp, New Romeo's Camp) and Site 2 located along the Kamoa River (October 21-26, 2006; Kamoa River). Sampling consisted of: (i) intensive hand collecting in leaf litter, rotten logs, fallen trees, and vegetation, and (ii) 60 leaf- and 40 wood-litter samples utilizing maxi-Winkler litter extractors. Each transect consisted of ten separate collections of sifted litter, six liters each, each sample collected separately from the others, and each sample consisting of litter from one or more microhabitats. Leaf-litter samples were taken from leaf litter (including small twigs and branches), whereas wood-litter samples were taken exclusively from rotten and decaying logs. Sampling followed a modification of the well-known and extensively utilized ALL protocol (Agosti et al. 2000). Sifted litter samples were suspended in mesh bags within the maxi-Winkler extractors for 48 hours. Collected specimens were preserved in 95% ethanol for subsequent sorting and identification in the lab upon return to the United States.

RESULTS

Because several years' time is necessary for sorting and preparing the estimated 25,000 specimens collected, we are not yet able to provide a detailed report on the results. Preliminary results of the hand collecting are as follows: We collected a total of 34 ant genera representing 9 subfamilies (of 21 subfamilies currently defined for the family Formicidae (see Appendix 1)). Site 1 contained the larger number of genera (33), whereas Site 2 contained 22 genera. Both sites shared 22 genera out of the total of 34 collected. Site 1 contained a higher number of exclusive ant genera (11), i.e., genera not shared with Site 2, whereas Site 2 contained only one of the non-shared genera.

GENERAL IMPRESSIONS

Based on the ant fauna, both sites are minimally impacted by humans. We found no evidence that human activities such as hunting, fishing, logging, or mining had any noticeable effect on the ant fauna. No invasive ant species were encountered, whereas such species are regularly encountered in human-disturbed habitats. For example, Akuthupono, the old Wai-Wai village, contained high concentrations of *Solenopsis* ("fire ant") species commonly encountered in human-disturbed habitats. Large *Atta* nests are frequently encountered in disturbed habitats, whereas in undisturbed forests they are relatively rare. Such rarity was encountered at both RAP sites. Our preliminary impressions include:

- The genera *Pheidole*, *Crematogaster*, *Dolichoderus*, and *Camponotus* appear to be the most conspicuous members of the ant fauna at both sites.

- The genus *Paraponera* and its only species, *P. clavata*, was found at Site 1 but not at Site 2.

- *Atta* sp. were collected at both sites, but were not common.

- *Mycetarotes* cf. *acutus* was collected at Site 1. This represents a surprising and significant range extension for this fungus-growing genus, previously known only from Amazonian Brazil and Argentina.

- Comparisons between sites suggest that Site 1, Acarai Mountains, contains a more diverse ant fauna than Site 2, Kamoa River. However, it is important to note that there was a greater sampling effort by the ant team at Site 1 than at Site 2. Once we complete thorough quantitative analyses of the data we will have an improved ability to compare diversity between the sites.

CONSERVATION RECOMMENDATIONS

No recommendations can be made at this stage. Such recommendations can only be formulated once the litter samples are fully analyzed.

REFERENCES

Agosti, D., J.D. Majer, L.E. Alonso and T. R. Schultz (eds.). 2000. Ants: Standard Methods for Measuring and Monitoring Biological Diversity. Smithsonian Institution Press. Washington, D.C.

Fernández, F. and S. Sendoya. 2004. List of Neotropical Ants. Biota Colombiana 5(1): 1-93.

Kaspari, M. and J.D. Majer. 2000. Using ants to monitor environmental change. *In:* D. Agosti, J. Majer, L. E. Alonso and T. R. Schultz (eds.). Ants, Standard Methods for Measuring and Monitoring Biodiversity. Washington, DC: Smithsonian Institution Press.

Kempf, W.W. 1972. Catalago abreviado das formigas da regiao Neotropical. Studia Entomologica 15: 3-344.

LaPolla, J.S., T. Suman, J. Sosa-Calvo and T.R. Schultz. 2007. Leaf litter ants of Guyana. Biodiversity and Conservation 16: 491-510.

Weber, N.A. 1946. The biology of the fungus-growing ants. Part IX. The British Guiana species. Revista de Entomologia (Rio de Janeiro) 17: 114-172.

Wheeler, W.M. 1916. Ants collected in British Guiana by the expedition of the American Museum of Natural History during 1911. Bulletin of the American Museum of Natural History 35: 1-14.

Wheeler, W.M. 1918. Ants collected in British Guiana by Mr. C. William Beebe. Journal of the New York Entomological Society 26: 23-28.

Chapter 3

Dung Beetles of Southern Guyana - a preliminary survey

Christopher J. Marshall

INTRODUCTION

A number of key features of dung beetles (Coleoptera: Scarabaeidae: Scarabaeinae) are largely responsible for their being used as biodiversity indicators. They are distributed on all of the continents and can be found from the equator to subarctic regions – inhabiting an array of terrestrial ecosystems. Diversity is often highest in tropical rain forests but temperate forests, grasslands and even xeric or high elevation sites can support diverse dung beetle assemblages. Because most dung beetles rely on dung from vertebrates, their abundance and diversity often reflects that of local vertebrates. A number of scarabaeine beetles are associated with non-dung food sources (e.g., fungi, millipedes or termites). These species are rarely collected in dung traps – thus it is important to note that this survey will reflect only a portion (albeit the majority) of the full scarabaeine beetle fauna for this region. Additionally, the dung beetles themselves support a vast assortment of symbionts, primarily mites, nematodes and fungi and as such are an indirect measure of these taxa.

One important characteristic of dung beetles that sets them apart from other potential biodiversity indicator taxa is the ease (and rapidity) with which they can be collected using baited traps. Previous studies have shown that a large percentage of the species (alpha diversity) for a region can be surveyed in a relatively few number of days. However, scarabaeine beetles do show some seasonality in their abundance and thus sampling across several seasons can be expected to yield a better overall diversity measure.

The ability to identify the dung beetles collected on this study varies depending on the genera. Some new world genera can be readily identified to species, however, some groups (e.g., *Dichotomius* or *Deltochilum*) are greatly in need of modern revisions and many species are poorly described or remain unnamed. Other groups (e.g., *Phanaeus, Coprophanaeus, Canthon, Ontherus,* etc.) are better known and species, including undescribed species, can be more easily determined using modern keys.

This report summarizes preliminary results of dung beetle collecting at two sites in southern Guyana, near the border of Brazil. These data will later be compared with preexisting data from other sites across the Guayana Shield and elsewhere in South and Central America.

METHODS

Study site

The Acarai Mountains are seasonally wet, forested, low to mid- altitude (<1500 m) uplands located in the southern part of Guyana. The Acarai Mountain range lies along the border shared between Guyana and Brazil, and is one of four mountain ranges in Guyana. Two important Guyanese rivers, the Essequibo (the longest river in the country and the third largest river in South America) and the Courantyne, originate in the Acarai Mountains. The Acarai Mountains are actually one part of a larger range that extends into the Wassarai Mountains to the north and east. North of the Acarai Mountains is the Sipu River, a western tributary of the Essequibo River. Along the river 200-400 m elevation, the forest resembles classic low elevation Amazonian forest. It is clear from the vegetation and sandy soil that this habitat floods annually. Terra firme forest is present above 400 m and continues up to elevations approaching 1200 m. At Site 1 (Acarai Mountains), dung beetle traps were set at several elevations; approximately: 350 m, 500 m, 800 m, and 1000 m.

The second site (Site 2, Kamoa Mountains) was located along the Kamoa River (another western tributary to the Essequibo) which is further north and runs directly south of the Kamoa Mountains. Site 2 was not in the Kamoa Mountains, but east of the range. Only one transect was possible at this site, at approximately 500 m elevation in terra firme forest.

Field methods

Dung beetles were sampled at several sites in the Acarai Mountains between 6-19 October 2006. From 21-27 October 2006, the site on the Kamoa River was sampled. Sampling consisted of: (i) baited pitfall traps, (ii) hand sampling, and (iii) light trapping. Pitfall trapping was conducted along an approximately linear transect and included 10 pitfall traps, placed 50 m apart. Each trap consisted of a 32-ounce polyethylene cup, sunk into the ground up to its lip and partially filled with water. Over each trap was suspended a cheese-cloth enwrapped mass of dung (~30 g). A leaf was then placed over the entire trap to repel rain. The traps were emptied and rebaited after 24 hours and left in the field for a total of 48 hours. Additional traps, baited with other baits, were also set to retrieve any species not associated with dung. Dung beetles seen perched on leaves while hiking in the forest were placed into vials. Lastly, light traps were set up (UV or Mercury Vapor) to sample several non-dung scarabaeine beetles (aphodiine dung beetles) that can be attracted in this manner.

Sampling largely follows procedures set forth by the ScarabNet group, a consortium of dung beetle biologists working around the world to better understand dung beetle diversity, and will allow these data to be compared to similar datasets. All specimens collected were preserved in 95% ethanol for subsequent sorting and identification.

RESULTS

A significant portion of the material collected still remains to be sorted and prepared in the lab — and thus species-level results from this sampling method are not reported here. Some genera have been mounted, so a preliminary account of the genera is possible. These initial data indicate that Site 1 (Acarai Mountains) supports a more diverse assemblage of species and genera than does Site 2 (Kamoa River). At Site 1, the following genera were observed: *Deltochilum*, *Ateuchus*, *Dichotomius*, *Ontherus*, *Canthon*, *Eurysternus*, *Oxysternon*, *Phanaeus*, and *Cryptocanthon*. At Site 2, no *Phanaeus*, *Deltochilum* or *Cryptocanthon* have yet to be found in the samples, but it is possible that they are present in the unprepared samples. The multiple elevations assessed at Site 1 appear to show different scarabaeine fauna although a few species, such

as *Oxysternon festivum*, were common at all sites, irrespective of elevation.

Based on the dung beetles so far identified, the habitat associated with these forested sites appears unaffected by human activity. Evidence of human trails (perhaps hunting/fishing/migration) was minimal.

CONSERVATION RECOMMENDATIONS

Given the relatively pristine nature of this region, the precautions necessary for conserving the native dung beetle fauna largely stem from potential threats of development. Agriculture and livestock cultivation are the clearest threats. Although many dung beetle species prefer disturbed agricultural systems, the regional endemics would not be expected to survive as they are particularly sensitive to micro-climatic variables such as humidity, tree cover, temperature, and soil type.

Hunting, given its potentially negative impact on mammalian and bird populations, could also negatively affect dung beetles. However, at this time, the hunting by the native Wai-Wai community does not appear to present itself as a major threat. If their population expands significantly or they begin to hunt commercially, this might change.

Human habitation and the associated waste management systems have also been suggested as potential impacts to local dung beetle communities. A large quantity of exposed human or animal waste would undoubtedly attract large numbers of dung beetles; however it is somewhat difficult to imagine this threat in the absence of far more threatening development (e.g., agriculture, habitation and deforestation).

Chapter 4

Water quality at selected sites in the Konashen COCA, Southern Guyana

Maya Trotz

INTRODUCTION

Water is an essential natural resource and its quality and quantity generally guide management plans for its use. Water quality data indicates the existing health of a water body and highlights any current or potential threats to the water body. The data collected from the rapid assessment (RAP) can be used as a baseline for a water quality monitoring program for the 2500 square mile Konashen area, which supports a population of approximately 200 people, and diverse plant and animal life. Water quality is also an appropriate indicator for RAP surveys because it provides information on conditions required for some taxonomic groups studied by other researchers. Water quality plays a major role in human health, and an important aspect of this assessment was to test water quality in areas that the local Wai-Wai community regularly depends on. These include potable water in Masakenari village and water in two main fishing grounds, Wanyakoko and Kanaperu. As owners and managers of the areas, the well-being of the local community is critical to the long-term sustainability of the area. This assessment did not include water resources measurements which are just as important as water quality data for the development of a water management and monitoring program for the area.

STUDY SITES

The five main sites sampled in the Konashen district were Sipu (SR), Acarai (AM), Kamoa (KR), Essequibo (ER), and Masakenari (MA). The complete list of sampling sites is included in Table 4.1 and a few are shown in Figure 4.1.

Sipu: The Sipu site was located on the Sipu River which is a tributary of the Essequibo River. There were no human or industrial sources of pollution evident at this site and dead trees were scattered throughout the river.

Acarai: The Acarai site was located below the Acarai Mountains adjacent to a creek that empties into the Sipu River. This creek was underlain with sand and rock and showed no human or industrial signs of pollution. Oral recollection from the Wai-Wai community signaled the presence of an illegal small-scale gold mining operation in the area around 1990. The two small creeks and a small stagnant pond sampled around the area were well shaded, covered with leaves, and less than a meter deep at the time.

Kamoa: The Kamoa site was located next to the Kamoa River, a tributary of the Essequibo River which was downstream of, and larger than the Sipu River. There were no human or industrial sources of pollution evident at this site and dead trees were scattered throughout the river.

Essequibo: The major river running through the Konashen District is the Essequibo which empties into the Atlantic Ocean. Sampling was conducted between the mouth of the Sipu River and the Kanaperu fishing ground. The village moved from Akuthopono in 2000 following a massive flood.

Table 4.1. Water quality sampling sites.

Site	Elev. (m)	Latitude hddd.ddddd°	Longitude hddd.ddddd°	Description
Sipu River				
GR-SR-01	249.6	01.25.595	- 058.57.044	
GR-SR-02	249.6	01.25.558	- 058.56.958	
GR-SR-03	236.2	01.42.293	- 058.95.154	
GR-SR-04	257.6	01.42.340	- 058.95.202	Black water creek off of Sipu River
GR-SR-05	237.7	01.38.990	- 058.94.486	Black water creek between Acarai and Sipu
GR-SR-06	257.9	01.43.072	- 058.92.941	Intersection of Acarai creek and Sipu River
Acarai Creek				
GR-AM-01	255.4	01.42.180	- 058.95.221	Small creek close to Acarai
GR-AM-02				Creek close to Acarai
GR-AM-03				Acarai creek at bridge between Sipu and Acarai
GR-AM-04	256.6	01.38.989	- 058.94.489	Acarai creek next to camp site
GR-AM-05	279.5	01.38.994	- 058.94.500	Isolated pool next to Acarai creek
Kamoa River				
GR-KR-01	224.6	01.53.179	- 058.82.983	Upstream of Kamoa River site
GR-KR-02	207.3	01.53.189	- 058.82.967	Northern side of the Kamoa River
GR-KR-03				Upstream of Kamoa River site
GR-KR-04				Upstream of Kamoa River site
GR-KR-05				Kamoa River site
GR-KR-06	241.4	01.53.427	- 058.82.692	Creek off of the Kamoa River
GR-KR-07				Creek off of the Kamoa River
GR-KR-08				Creek off of the Kamoa River
GR-KR-09	249	01.53.135	- 058.82.226	
GR-KR-12	235	01.53.193	- 058.81.922	Creek off of the Kamoa River
GR-KR-12b	235	01.53.193	- 058.81.922	Creek off of the Kamoa River
GR-KR-13	227.4	01.52.729	- 058.73.961	
GR-KR-14	226.2	01.52.840	- 058.73.535	Intersection of Essequibo and Kamoa rivers
Essequibo				
GR-ER-01	221.9	01.65.857	- 058.62.648	Akuthopono landing
GR-ER-02				
GR-ER-03	213.4	01.64.913	- 058.62.228	Essequibo River between Masakenari and Akuthopono
GR-ER-04	217.9	01.64.537	- 058.61.705	Essequibo River
GR-ER-05	212.4	01.63.899	- 058.62.064	Essequibo River
GR-ER-06	224.9	01.63.596	- 058.62.860	Essequibo River
GR-ER-07	225.2	01.63.232	- 058.62.268	Essequibo River
GR-ER-08	222.8	01.62.764	- 058.63.016	Essequibo River
GR-ER-09	224.9	01.62.957	- 058.62.429	Western side of the bank
GR-ER-10	221.6	01.62.963	- 058.62.436	Middle of the river
GR-ER-11	237.7	01.62.976	- 058.62.447	Eastern side of the bank

Site	Elev. (m)	Latitude hddd.ddddd°	Longitude hddd.ddddd°	Description
GR-ER-12	212.4	01.64.733	- 058.61.826	Rapids between Masakenari and Akuthopono
GR-ER-13	236.2	01.70.889	- 058.62.016	Kanaperu fishing ground
GR-ER-14	230.1	01.70.889	- 058.62.016	Kanaperu fishing ground
GR-ER-15	228.3	01.68.172	- 058.62.938	Wanyakoko fishing ground, middle of pond
GR-ER-16	232.9	01.68.102	- 058.62.934	Wanyakoko fishing ground, side of pond
GR-ER-17	222.2	01.43.083	- 058.92.948	Intersection of Sipu River and Essequibo River
GR-ER-18	242	01.42.147	- 058.80.210	
GR-ER-19	230.7	01.48.091	- 058.78.892	Upstream of the Kamoa River
Masakenari				
GR-MA-01				Black water creek, upstream of bathing
GR-MA-02				Black water creek, downstream of bathing
GR-MA-03	256	01.62.896	- 058.63.527	Pipe 1
GR-MA-04				Pipe 2, next to school
GR-MA-05				Pipe 3
GR-MA-06				Pipe 4
GR-MA-07	242	01.63.167	- 058.63.471	Village well
GR-MA-08	236.2	01.65.148	- 058.63.627	Palm swamp between Akuthopono and Masakenari

Masakenari: The village has approximately 200 residents and the main supply of potable water comes from a 3.6 m deep well, located at the elevation of 242 m, somewhat below the village which starts at approximately 256 m. Well water is pumped by a solar powered pump to three plastic vats, where they are then piped to four pipes located throughout the village. The well is 1.5 m wide, and made of concrete with a loosely packed clay brick bottom. At the time of this assessment the water level was at 239 m (~ 1 m from the well bottom). The community uses latrines, most of which are located above the well. A creek runs through the community, and is used for bathing and washing. The creek was scattered with dead trees and the only sign of garbage was an empty paint can in the middle of a point approximately 50 m upstream of bathing. Masakenari was approximately 5.8 km upstream of Akuthopono, the site of the old village. In 2000 Akuthopono experienced massive flooding with water levels rising over 9 m from current levels, an estimation based on tree markings made by the Wai-Wai community.

METHODS

A HYDROLAB Quanta multi-sensing system was used to conduct water quality tests in the field (pH, conductivity (mS/cm), turbidity (NTU), and DO (mg/L)). The instrument was calibrated using pH 4 and 7 buffers, 0 and 20 NTU standards, and a 1.412 mS/cm conductivity standard. 100% DO_{sat} was calibrated using MQ water that had been equilibrated with the atmosphere. Measurements were taken as a function of depth (0, 0.8 m and the bottom) at various locations along the center of the Essequibo River, Sipu River, Acarai creek and Komoa River of Guyana. Sampling was also done in creeks and ponds around Sipu, Acarai, Kamoa, Akuthopono and Masakenari. In Masakenari, well water was sampled at the four village pipes and surface water was sampled upstream and downstream of a creek used for domestic purposes.

Water samples were taken from 1 cm below the surface. Alkalinity measurements were made within 24 hours by titrating 40 mls of samples with 0.02 N H_2SO_4 to pH 4.3 and a methyl orange end point. Water samples were also acidified with ultra pure nitric acid to give a 10% acid solution. Some samples were also filtered using a 0.2 μm PES filter (Nalgene) and acidified with nitric acid. All acidified samples were stored for elemental analysis. Samples for mercury analysis were collected from the fishing areas Waynakoko and Kanaperu. This was done in ultra pure HDPE containers that were doubly bagged in plastic zippered storage bags and by two people to reduce sample contamination.

Sediment samples were taken from the banks just below the water surface or in the case of shallow creeks, from the bottom. These were placed in plastic storage bags and stored in a freezer for further analysis. GPS measurements were taken using a Garmin Etrex with a reference point of Prov S Am '56.

Table 4.2. Summary of surface water quality results from large water bodies.

	Essequibo River	Sipu River	Acarai Creek	Kamoa River
pH	5.11 – 6.53	5.49 – 6.24	5.24 – 5.49	5.91 – 6.12
DO (mg/L)	4.26 – 8.25	6.45 – 7.43	7.50 – 8.25	5.88 – 6.91
Turbidity (NTU)	0 – 12.1	5.0 – 19.7	0 – 2.5	23.7 – 43
Alkalinity (mg/L CaCO$_3$)	12.5 – 13.5	10 – 12.5	–	7.5 – 18.75
Conductivity (mS/cm)	< 0.02	< 0.02	< 0.02	< 0.02

Table 4.3. Summary of surface water quality of selected small ponds and creeks.

	Palm Swamp bet Masakenari and Akuthopono (GR-MA-08)	Creek off of Sipu River (GR-SR-04)	Isolated pool adjacent to Acarai Creek (GR-AM-05)	Creek off of Kamoa River (GR-KR-08)
pH	4.81	5.9	4.83	4.89
DO (mg/L)	2.85	4.73	3.19	6.15
Turbidity (NTU)	23.7	19.4	27.1	7.3
Alkalinity (mg/L CaCO$_3$)	–	27.5	–	10
Conductivity (mS/cm)	0.016	0.03	0.012	0.011

Table 4.4. Summary of water quality results from Masakenari drinking water.

	Well (0 m depth)	Well (1 m depth)	Pipe 1	Pipe 2	Pipe 3	Pipe 4
pH	5.27	6.09	6.49	6.42	6.55	6.47
DO (mg/L)	3.22	2.57	5.63	5.06	4.89	5.32
Turbidity (NTU)	8.5	–	0.1	0	1.4	0.5
Conductivity (mS/cm)	0.043	0.043	0.049	0.046	0.055	0.047

RESULTS

The pH of sampled water ranged from 4.74 to 6.24, with the lower pH values obtained in isolated ponds and small creeks, and the higher pH readings seen in the rivers. Alkalinity values ranged from 7.5 to 27.5 mg/L CaCO$_3$. Dissolved oxygen levels ranged from 2.85 to 8.25 mg/L, and were generally lower in isolated ponds and small creeks than in the rivers. Conductivity was below 0.02 mS/cm for most of the waters sampled and was between 0.04 and 0.06 mS/cm in Masakenari village. Table 4.2 summarizes data obtained from the major creek and rivers assessed, and Table 4.3 summarizes data from a subset of the small creeks or isolated pools at various study sites.

Table 4.4 summarizes water quality data obtained in the main drinking water sources for Masakenari village. The pH of the water from the four village pipes was between 6.4 and 6.5, whilst that of 1 m of well water was between 5.3 and 6. Dissolved oxygen levels of the well were below 3.2 mg/L, and at the pipes ranged from 4.9 to 5.6 mg/L. Turbidity of the well and pipe water was below 10 NTU, and conductivity was between 0.04 and 0.06 mS/cm. USEPA/WHO standards for drinking water pH and turbidity are 6.5 – 8.5 and less than 5 NTU, respectively. No standards exist for conductivity, alkalinity or dissolved oxygen.

CONSERVATION RECOMMENDATIONS

Basic water quality data and observations show that the main rivers and creeks of Konashen are free of human or industrial pollution. The pH values of the majority of creeks, rivers and isolated pools are similar to those observed in the Amazon basin, but are lower than the drinking water standards of the WHO or USEPA. Those of the village wells are close to the minimum requirement of 6.5 pH units. Similarly, turbidity levels of some of the rivers were higher than the drinking water standards of 5 NTU. These untreated surface waters are used for drinking and cooking for transient camps. Even though conductivity and hence total dissolved solids concentrations are low, further sample analysis will provide information on the concentration of heavy metals which have extremely low drinking water standards.

The rivers and creeks are used for domestic purposes and only in Masakenari is drinking water obtained from a well. Well water should be monitored on a consistent basis as the well sits downstream of the village garbage holes and latrines, which are unlined. In Akuthopono, water is collected from the river, and garbage is dumped in a hole which was used by the village as a well until the year 2000. This activity could potentially affect the groundwater quality and plans should be made to provide clean drinking water at the site if it is to be developed as an income-generating ecotourism visitor center.

A water quality monitoring program should be established in the area to expand and continue the work begun during this rapid assessment. Such a program should be conducted on a quarterly basis at selected sampling sites along the Essequibo River (e.g. GR-KR-14, GR-ER-17, GR-ER-01, GR-ER-14, GR-ER-15), and in the village (GR-MA-01 – GR-MA-07). This quarterly monitoring should provide two samples each, during the dry and wet seasons.

Water quality monitoring should be extended to include microbial analysis in the more populated area. Sites used to collect aquatic species as well as any new sites identified by the Wai-Wai community should also be monitored for water quality on at least an annual basis.

Plans should be made to quantify the water resources in the area, especially since the area experiences high levels of rainfall and is inundated for large periods of the year. This information will also assist with safer plans for water and sanitation in Masakenari and Akuthopono.

Figure 4.1. Water quality sampling sites in the Konashen COCA

GR-MA-02: Downstream of bathing area at the creek in Masakenari

GR-SR-04: Creek off of the Sipu River

GR-AM-03: Acarai creek

GR-AM-04: Acarai creek

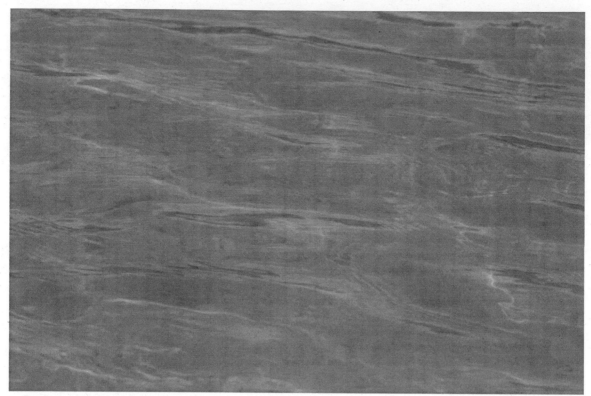

GR-ER-13: Kanaperu fishing ground

GR-ER-01: Akuthopono Landing

Chapter 5

Aquatic Biota: Fishes, Decapod Crustaceans and Mollusks of the Upper Essequibo Basin (Konashen COCA), Southern Guyana

Carlos A. Lasso, Jamie Hernández-Acevedo, Eustace Alexander, Josefa C. Señaris, Lina Mesa, Hector Samudio, Julián Mora-Day, Celio Magalhaes, Antoni Shushu, Elisha Mauruwanaru and Romel Shoni

SUMMARY

During the period from October 15 -26, 2006, a rapid assessment of the aquatic ecosystems of the Acarai Mountains and Sipu, Kamoa and Essequibo rivers upstream from the Amaci Falls, Konashen Indigenous District of Southern Guyana was conducted. We studied fishes, crustaceans and mollusks at 18 sampling stations within five focal areas: 1) Focal Area 1 – Sipu River; 2) Focal Area 2 – Acarai Mountains; 3) Focal Area 3 – Kamoa River; 4) Focal Area 4 – Wanakoko Lake/Essequibo River; and 5) Focal Area 5 – Essequibo River at Akuthopono and Masakenari Village. A total of 113 species of fish were identified, representing six orders and 27 families. The order Characiformes (tetras, piranhas, etc.) with 61 species (51.7%) was the most diverse, followed by Siluriformes (catfishes) with 32 species (27.1%), Perciformes (cichlids, drums) and Gymnotiformes (electric or knife fishes) with nine species each (15.3% respectively), and Cyprinodontiformes (killifishes) and Synbranchiformes (eels), both with one species (0.8%, respectively). Family Characidae contributed the most species, with 31 species collected (27.4%), followed by Loricariidae with 13 species (11.5%); Cichlidae with 8 species (7.1%); Crenuchidae, Curimatidae, Anostomidae and Heptapteridae with 5 species each (4.4%, respectively), and Auchenipteridae, Callichthyidae and Erythrinidae, with 4 species each (3.5% respectively). The 17 remaining families represented a combined total of 29 species (25.7%). Focal Area 5 exhibited the highest species richness, with 53 species of the 113 identified (46.9%), followed by Focal Areas 1 and 3, with 48 and 45 species, respectively (42.58% and 39.8%), while Focal Areas 4 and 2 had 33 and 32 species, respectively (29.2% and 28.3%). According to the distribution of fish species, and based on the similarity index and physicochemical variables, Focal Areas 1 and 3 exhibited the highest similarity (0.67), and can be viewed as possessing similar ichthyological communities. The remaining Focal Areas exhibited lower values, between 0.4 and 0.26, and are therefore considered to be of moderate similarity. Nearly half of the fish species we recorded are considered important subsistence fish resources, 20% are of sport fishing interest and approximately 75% have ornamental value. Four species of fishes are considered likely to be new to science (*Hoplias* sp., *Ancistrus* sp., , *Rivulus* sp., and *Bujurquina* sp.). Ten species of aquatic macroinvertebrates were identified, belonging to three classes (Crustacea, Gastropoda, and Bivalvia), of which Crustacea was the most diverse, with three families. Of these, Pseudothelphusidae showed the highest richness, with four species, followed by Palaemonidae and Trichodactylidae with two species each. The classes Gastropoda (snails) and Bivalvia (mussels) were represented by one species each. The greatest species richness was found in Focal Areas 2 and 3, with five and six species of aquatic macroinvertebrates respectively, whilst three species were collected in each of the remaining focal areas, except for Focal Area 5 where four species were recorded.

INTRODUCTION

With 700 known fish species, Guyana is arguably the best studied country in the Guayana Shield from an ichthyological perspective, followed by French Guiana. However, within Guyana there is still a scarcity of information as many regions remain unstudied (Lasso et al. 2003). Initial studies were carried out by North American ichthyologist Carl Eigenmann at the turn of the century, and covered a large part of the Guyanese territory. He studied eight localities in the Lower Essequibo and published the results in a comprehensive summary in 1912. Much later, Watkins et al. (1997) and Hardman et al. (2002) collected again at the Lower Essequibo and compiled an updated study, compiling a list of nearly 400 species for the basin of the Lower Essequibo (Lasso 2002). Nevertheless, all these efforts were concentrated on the Lower Essequibo, while the Upper Essequibo remained virtually unstudied. In 2001, Conservation International conducted the second Rapid Assessment Program (RAP) expedition in the Eastern Kanuku Mountains, in the Lower Kwitaro River and the Upper Rewa River at Corona Falls. During this survey 113 species were documented (Mol 2002).

The present RAP expedition is the first comprehensive fish and crustacean investigation of the Acarai Mountains, Sipu, Kamoa and Essequibo rivers upstream from the Amaci Falls. These data are new as the aquatic biota of the Essequibo's headwaters have never been studied, and the waters never before characterized. The hydrochemistry (Chapter 4) and data on the aquatic fauna from this study, coupled with a mini-survey on fishing resources of Masakenari carried out by Alexander et al. (2005) constitute a significant contribution to the knowledge of the biodiversity of Guyana.

In addition, in 2002, the Fishes and Freshwater Ecology of the Guayana Shield Conservation Priorities Consensus recognized the Acarai Mountains as a region completely unexplored biologically, and emphasized the need for surveys in the area, deeming it a conservation priority (Lasso et al. 2003).

METHODS AND STUDY SITES

During the period from October 15-26, 2006 we surveyed 18 sampling stations within five focal areas (see Table 5.1):

Focal Area 1: Sipu River: six sampling stations (GR-SR-01 to GR-SR-05 and GR-SR-08).

Focal Area 2: Acarai Mountains: three sampling stations (GR-AM-06 to GR-AM-07a, b).

Focal Area 3: Kamoa River: four sampling stations (GR-KR-09 to GR-KR-12).

Focal Area 4: Wanakoko Lake/Essequibo River: one sampling station (GR-WL-13).

Focal Area 5: Essequibo River at Akuthopono and Masakenari Village: four sampling stations (GR-AR-14 a, b; GR-PF-15; GR-MAR-16).

Table 5.1. Localities studied during the 2006 RAP survey of the Acarai Mountains, Sipu, Kamoa and Essequibo rivers, Konashen Indigenous District of Southern Guyana.

CODE	Locality	Coordinates	Focal Area
GR-SR-01	Sipu River	1°25.558 N-58°56.958 W	
GR-SR-02	Sipu River	1°25.558 N-58°56.958 W	
GR-SR-03	Sipu River	1°42.293 N-58°95154 W	AF 1
GR-SR-04	Sipu River - small creek	1°42.340 N-58°95202 W	
GR-SR-05	Sipu River - isolated pool	1°25`05.9`` N-58°57`12.4`` W	
GR-AM-06	Acarai creek	1°42180 N-58°95221 W	
GR-AM-07a	Acarai creek marginal pool	1°42180 N-58°95221 W	AF 2
GR-AM-07b	Acarai creek	1°42180 N-58°95221 W	
GR-SR-08	Sipu River - small creek	1°38990 N-58°94486 W	AF 1
GR-KR-09	Kamoa River	1°51`51.1``N-58°49`41.9``W	
GR-KR-10	Kamoa River - small creek	1°31`46.5``N-58°49`14.7``W	AF 3
GR-KR-11	Kamoa River - small creek	1°31`48.6``N-58°48`34.5``W	
GR-KR-12	Kamoa River - small creek	1°31`42.3``N-58°49`14`W	
GR-WL-13	Wanakoko Lake - Essequibo River	1°40`41.2``N-58°37`50``W	AF 4
GR-AR-14a	Essequibo River - palm swamp Akothopono	1°65148 N-58°62367 W	
GR-AR-14b	Essequibo River - Akuthopono rocks	1°39`02.4``N-58°37`40.5``W	AF 5
GR-PF-15	Essequibo River - Akuthopono forest	1°39`02.4``N-58°37`40.5``W	
GR-MAR-16	Essequibo River - Akuthopono rapids	1°34`08.8``N-58°38`48.9``W	

Fishes and aquatic invertebrates (crustaceans and mollusks) were collected during both night and day using several
methods: two gill nets were put out daily between the hours
of 5:30 and 8:30 and between 17:00 and 19:00. In the
small creeks we employed a 2 m seine net (height = 1.1 m,
mesh size = 1 mm). In addition, we used 10 minnow traps
daily to collect small fish and crustaceans. The fish team also
conducted manual collecting using a dip net and, in the
Essequibo rapids, medium-sized fishes were captured using
a cast net. On one occasion (Acarai creek), we employed a
traditional Wai-Wai technique and used a natural ichthyo-
cide extracted from lianas of hiari (*Derris elliptica*), a plant
native to Guyana. We sampled a variety of different habitat
types including the main channels of rivers (open waters,
littoral or river banks, pocket waters with rocks and rapids,
e.g. Sipu, Kamoa and Essequibo rivers), side pools (standing waters of the Essequibo River at Wanakoko Lake), small
lowland creeks (clear and black waters); mountain creeks
(clear waters, e.g. foothills of Acarai Mountains), and palm
swamps and seasonally dry ponds (e.g. flooded forests of
lower Essequibo River near Akuthopono). We surveyed all
encountered microhabitats e.g., riffles, pools, leaf litter and
woody debris. In addition, we recorded underwater observations. Biophysical characteristics (general description), hydrochemical traits and georeference points were recorded for
all localities sampled.

Laboratory work

Fishes were preserved in 10% formalin and later transferred
to 70% ethanol. Samples were deposited in the Center for
the Study of Biological Diversity of the University of Guyana, Georgetown, and a small reference collection was taken
for identification to the Museo de Historia Natural La Salle,
Caracas (Venezuela).

In order to establish the level of similarity of fish communities between localities, the Simpson Index of similarity
was used ($RN2 = 100 (s) / N2$), where s is the number of
species shared between both subregions or localities, and $N2$
is the number of species in the subregion or locality with the
lowest richness. Principal component and cluster analysis
were also done, using the statistical package PAST (Hammer
et al. 2001) to graphically group the localities.

RESULTS AND DISCUSSION

Fishes

Composition and species richness

During the RAP expedition to the Konashen COCA Southern Guyana, a total of 2651 specimens belonging to 113
species in six orders and 27 families were collected (Appendix 2). The order Characiformes (tetras, piranhas, etc.), with
61 species (51.7%), was the most diverse, followed by Siluriformes (catfishes), with 32 species (27.1%), Perciformes
(cichlids, drums) and Gymnotiformes (electric or knife
fishes), with nine species each (15.3% respectively), and finally Cyprinodontiformes (killifishes) and Synbranchiformes
(eels), both with one species (0.8% respectively) (Figure
5.1). Family Characidae contributed the most species with
31 species collected (27.4%), followed by Loricariidae with

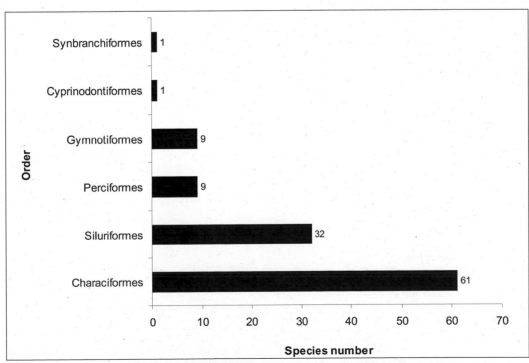

Figure 5.1. Richness of fish orders reported during the 2006 RAP survey of the Acarai Mountains, Sipu, Kamoa and
Essequibo rivers, Konashen Indigenous District of Southern Guyana.

13 species (11.5%); Cichlidae with 8 species (7.1%); Crenuchidae, Curimatidae, Anostomidae and Heptapteridae with 5 species each (4.4% respectively), and Auchenipteridae, Callichthyidae and Erythrinidae with 4 species each (3.5%, respectively). The 17 remaining families represent a combined total of 29 species (25.7% in total) (Figure 5.2).

Results for Focal Areas

In order to make comparisons of species richness, the study area was divided into five focal areas: Focal Area 1: Sipu River (SR), Focal Area 2: Acarai Mountains (AM), Focal Area 3: Kamoa River (KR), Focal Area 4: Wanakoko Lake (WL)/Essequibo River (AR, PF and MAR), and Focal Area 5: Essequibo River at Akuthopono and Masakenari Village. Focal Area 5 was found to exhibit the highest species richness, with 53 of the 113 species identified (46.9%), followed by Focal Areas 1 and 3, with 48 and 45 species respectively (42.58% and 39.8%), while Focal Areas 4 and 2 exhibited 33 and 32 species respectively (29.2% and 28.3%) (Table 5.2, Appendix 2).

Taking into account the distribution of taxa, and based on Simpson's Index of similarity, Focal Areas 1 and 3 were shown to possess the highest similarity (0.67) and can be considered to have equal, or at least most similar, ichthyological composition. The other Focal Areas exhibit lower

values for this index, between 0.4 and 0.26, which can be considered as average similarity that diminishes as the value on the X-axis increases (Figure 5.3).

This distribution coincides with the behavior of physicochemical variables (Table 5.3). In the principal component analysis, the two primary ordination axes explained 89.67% of variation in the data, furthermore the variables pH and temperature were highly positively correlated (0.829), as were pH and conductivity (0.706). Figure 5.4 shows that Focal areas 4 and 5 (AF-4 and AF-5) are closely related with respect to water temperature since they possess similar average values; with respect to pH, Focal Area 4 exhibited the highest value, followed by Focal Areas 1 and 5. This is represented clearly in the graph, and is indicated by proximity of each focal area to the vector for pH. Conductivity exhibited highest values in Focal Areas 1 and 4. Focal Areas 1 and 3 exhibited highest values for dissolved oxygen. Focal Area 2 was found to be furthest from all measured vectors due to the low values recorded in the physicochemical variables of interest. Focal Areas 1 and 3 exhibited high correlation in the bi-plot since their physicochemical variables behave in a similar manner, in the same way the pairs consisting of Focal Areas 4-5 and 1-4 exhibited a medium correlation, whereas Focal Area 2 was found to be far away from these groups.

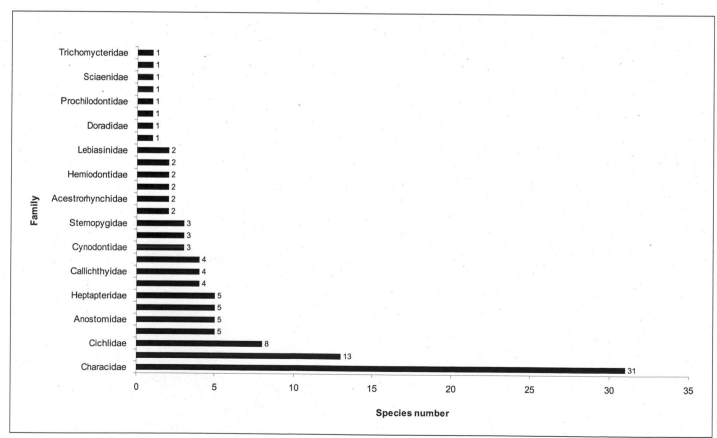

Figure 5.2. Richness of fish families reported during the 2006 RAP survey of the Acarai Mountains, Sipu, Kamoa and Essequibo rivers, Konashen Indigenous District of Southern Guyana.

Table 5.2. Fish species richness reported from focal areas evaluated during the 2006 RAP survey of the Acarai Mountains (AM), Sipu (SR), Kamoa (KR) and Essequibo rivers (ESSEQ), and Wanakoko Lake (WL), Konashen Indigenous District of Southern Guyana.

Order	Family	SR	AM	KR	WL	ESSEQ
Characiformes	Acestrorhynchidae	2	-	2	1	-
	Anostomidae	2	3	-	1	3
	Characidae	18	9	15	13	13
	Crenuchidae	2	3	2	-	4
	Curimatidae	3	-	2	2	2
	Cynodontidae	-	-	-	3	-
	Erythrinidae	3	1	1	1	4
	Hemiodontidae	1	-	1	1	2
	Lebiasinidae	2	-	2	2	2
	Parodontidae	-	1	-	-	1
	Prochilodontidae	-	-	1	-	-
Siluriformes	Auchenipteridae	1	-	3	1	1
	Callichthyidae	1	2	-	-	2
	Cetopsidae	2	1	-	-	-
	Doradidae	-	-	-	1	-
	Heptapteridae	1	1	3	-	2
	Loricariidae	4	7	1	-	9
	Pimelodidae	-	-	2	-	-
	Trichomycteridae	-	-	1	-	1
Cyprinodontiformes	Cyprinodontidae	-	1	-	-	-
Gymnotiformes	Gymnotidae	1	-	2	-	2
	Hypopomidae	-	-	2	-	2
	Rhamphichthyidae	-	-	1	-	-
	Sternopygidae	-	1	-	1	1
Synbranchiformes	Synbranchidae	1	-	1	-	-
Perciformes	Cichlidae	4	2	3	5	2
	Sciaenidae	-	-	-	1	-
Total		**48**	**32**	**45**	**33**	**53**

Focal Areas 1 and 2

Focal Areas 1 and 2 were completely pristine and well protected within the Konashen Indigenous District of Southern Guyana. The fish sampled in these focal areas were highly abundant, and of particular interest as subsistence resources. The high abundance, coupled with the large size of the fish that were collected and/or observed, indicate that the Sipu River and Acarai creek maintain intact populations of fish that have not been subject to exploitation. There are also significant populations of aimaras (*Hoplias macrophthalmus*) in both rivers. We sampled the main channel of the Sipu River (open waters and littoral or bank areas), small black water creeks, one dried pond of the Sipu River and a mountain clearwater creek (Acarai creek at the foothill of the Acarai Mountains). In the Sipu River, which included sampling in a flowing creek and an isolated pond, we observed high species richness. The Acarai creek is very important as its hydrochemical and other environmental characteristics clearly differentiate it from the other creeks and rivers studied. This is reflected in the composition of the aquatic biota, especially the fish. Many of the species collected are typical of the riffle microhabitat (e.g. Crenuchidae, Parodontidae, Loricariidae

and Hepapteridae). Of particular interest in the Acarai creek were the armored catfish (Family Loricariidae), tentatively assigned (pending further identification) to the genera *Ancistrus*. and could be endemic to the river basin and new to science. This could also be the case for the cichlid, *Bujurquina* sp., and the killifish, *Rivulus* sp., recorded in this study area.

Focal Area 3

Like the two preceding focal areas, Focal Area 3 (Kamoa River) is in pristine condition. The Kamoa River's fish composition and species richness are similar to that of the Sipu River, although somewhat different from the Acarai creek in species composition. In this region, the smaller fish dominated the clear and black water tributaries of the Kamoa River. We obtained a very representative sample of the creek's ichthyofauna. The richness of this system was lower but the vast majority of species were very tiny, associated with cryptic habitats and leaf litter. In the principal channel of the Kamoa River we also observed fish species of large size and in considerable abundance. There are important populations of aimaras (*Hoplias macrophthalmus*), which also indicate the presence of the tiger fish (*Pseudoplatystoma fasciatum*).

Table 5.3. Physico-chemical variables reported from focal areas evaluated during the 2006 RAP survey of the Acarai Mountains, Sipu, Kamoa and Essequibo rivers, Konashen Indigenous District of Southern Guyana.

Focal areas	Water temperature (°C)	pH	Dissolved oxygen (mg/l)	Conductivity (ms/cm)
AF1	25.30	5.85	6.11	0.02
AF2	23.90	5.02	4.74	0.01
AF3	25.13	5.44	6.32	0.01
AF4	28.10	6.10	5.70	0.02
AF5	28.23	5.83	4.74	0.01

Focal Area 4

Focal Area 4 (Wanakoko Lake) is not really a lake, but a large curvature of the main channel of the Essequibo River which is regularly fished by members of the Wai-Wai community. It is more similar to a side pool that is shallower than the river itself, and with calm waters. We recorded fish species typical of fast-moving, highly oxygenated river water (e.g. Acestrorhynchidae, Characidae, Erythrinidae), as well as species characteristic of slower, calmer river waters (e.g. Cichlidae, Curimatidae, Electrophoridae) in Wanakoko Lake. This region, according to preliminary results of a community-based fish mini-survey conducted by CI-Guyana (Alexander et al. 2005), is considered to be one of the four most important fishing areas in the Konashen Indigenous District.

Focal Area 5

In the Focal Area 5 (Essequibo River at Akuthopono and Masakenari Village) we studied four habitat types, which included the main channel of the Essequibo River (pocket water with numerous large rocks), one palm swamp, and one dried pond in the flooded forest of Akuthopono and the rapids of the Essequibo River between Akuthopono and Masakenari. We estimate that there were around 100 species in this area. The species numbers were low due to the cursory nature of our sampling of the habitats of the main channel (littoral area, banks and pocket waters). In the palm swamp we recorded some interesting species associated with standing water habitats, including some electric fish (Gymnotidae, Hypopomidae). In the dried pond of the flooded forest we observed a high abundance of *Hoplerythrinus unitaeniatus* (Erythrinidae), a species with aerial respiration, which allows it to tolerate the anoxic conditions of the pond. The rapids of the Essequibo River were better sampled than the pocket waters, especially the zones with rocks and aquatic plants of the family Podostemaceae (*Apinagia* sp. and *Mourera fluviatilis*), where the associated microichthyofauna is unique. In this habitat type, we collected many species of fish found only in this type of habitat (e.g. *Leporinus* spp., *Hemiodus* spp., *Rineloricaria platyura*, *Characidium* spp., *Melanocharacidium blennioides*, *Imparfinis* sp., etc.).

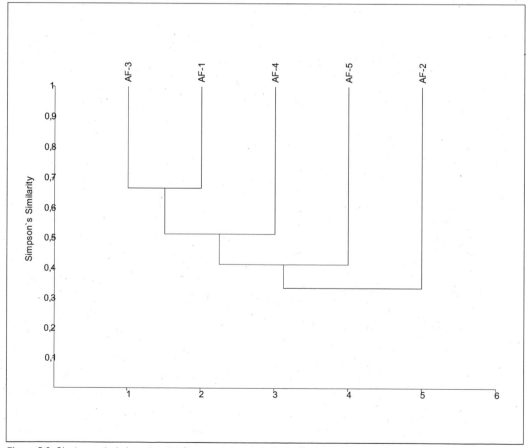

Figure 5.3. Cluster analysis based on the Simpson Index of similarity for the focal areas evaluated during the 2006 RAP survey of the Acarai Mountains, Sipu, Kamoa and Essequibo rivers, Konashen Indigenous District of Southern Guyana.

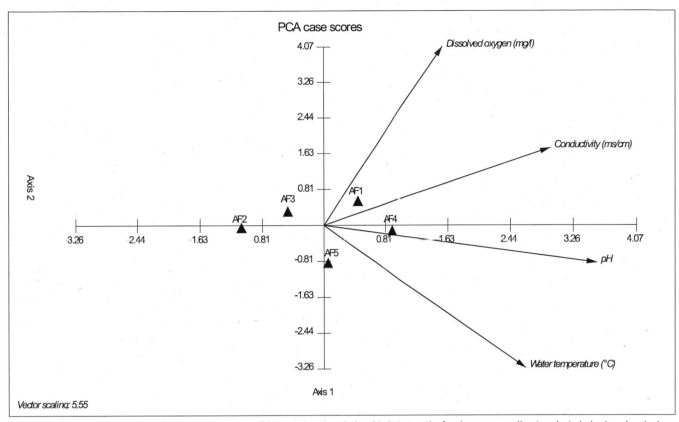

Figure 5.4. Biplot based on Principal Component Analysis (PCA) showing the relationship between the focal areas according to selected physico-chemical variables evaluated during the 2006 RAP survey of the Acarai Mountains, Sipu, Kamoa and Essequibo rivers, Konashen Indigenous District of Southern Guyana.

Results for each sampling station

18 sampling stations were evaluated during the RAP expedition to the Konashen COCA. Sampling results suggest that Wanakoko Lake (WL: Focal Area 4) possesses the highest species richness with 29.2% of the species collected, followed by Focal Area 2 (AM-07b), Focal Area 4 (PF-14b) and Focal Area 3 (KR-09), all with richness higher than 20%. The lower richness reported for Focal Area 2 or Acarai Mountains (AM-06 and AM-07a), with two species each, represents 1.77% of the total number of species identified (Appendix 2, Figure 5.5). The cluster analysis based on the Simpson Index of similarity for each of the sampling stations did not identify associations consistent with the distribution previously described for the focal areas (Figure 5.6). Thus, species richness alone is not a reliable variable for determining the type of relationship between the evaluated sampling stations.

Species Accumulation Curve

The species accumulation curve (Figure 5.7) provides evidence of the efficiency of sampling during the RAP expedition to the COCA. On the first day, 26 species were collected (representing 23% of the total captured), with a subsequent phased increase until day five, when no additional species were recorded. On day six the curve increased again with the addition of 13 species before stabilizing during day seven when no new species were added. During day eight the curve exhibited sustained growth until day ten with 10 more species added to the total collected during sampling.

The behavior of the curve demonstrates that sampling permitted the collection of a number of important species. However, the curve did not level out sufficiently to indicate that sampling effort was sufficient to record the majority of species present. The shape of the curve suggests that a number of species were not recorded in the sample, and that additional sampling of longer duration is necessary to record those species that were potentially excluded from the samples analyzed here.

Interesting Species

The fish team did not encounter any species currently recognized to be threatened (e.g. IUCN Red List, CITES, regionally or locally threatened). It is too early to determine accurately the endemism of the fish, mollusks and crustaceans that were collected since many of the species occur in the Lower Essequibo and are widely distributed throughout the Guianas. However, the samples are still being identified, and it is likely that some of the species collected will turn out to be endemic to the river basin of the Essequibo, especially members of the family Crenuchidae, and some of the Characidae, Hepapteridae, Cetopsidae, Rivulidae, and Cichlidae. Special attention should be given to the loricarid assigned tentatively to the genera *Ancistrus*; the killifish (*Rivulus* sp.), and the cichlid (*Bujurquina* sp.). It is important to note that these last four species, which are restricted to the Acarai Mountains, along with a species of aimara that lives only in the rapids of the Essequibo River (*Hoplias* cf. *malabaricus*), are thought to be new to science.

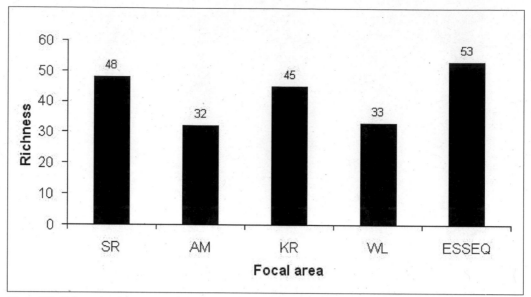

Figure 5.5. Fish species richness recorded in the focal areas evaluated during the 2006 RAP survey of the Acarai Mountains (AM), Sipu (SR), Kamoa (KR) and Essequibo (ESSEQ) rivers, and Wanakoko Lake (WL), Konashen Indigenous District of Southern Guyana.

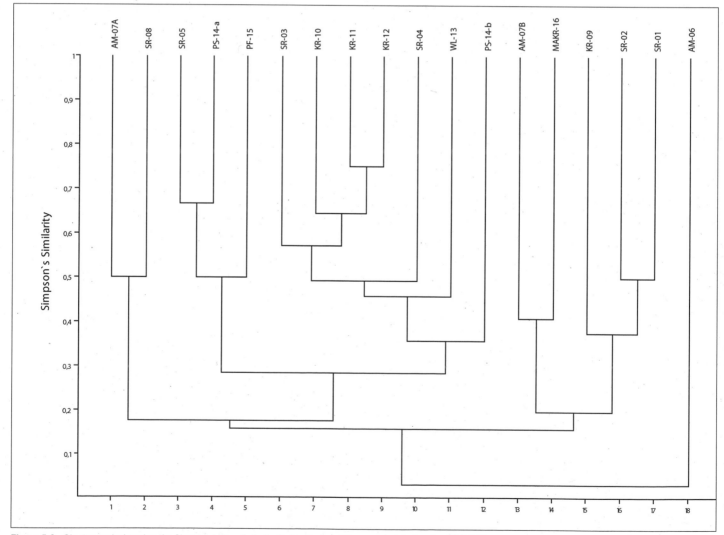

Figure 5.6. Cluster analysis using the Simpson Index of similarity for the localities sampled during the 2006 RAP survey of the Acarai Mountains, Sipu, Kamoa and Essequibo rivers, and Wanakoko Lake, Konashen Indigenous District of Southern Guyana.

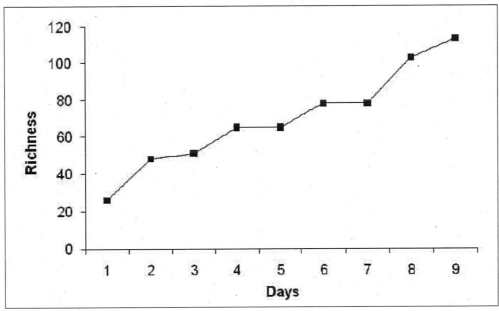

Figure 5.7. Accumulation curve for ichthyological species added to the overall species list per day of study during the 2006 RAP survey of the Acarai Mountains, Sipu, Kamoa and Essequibo rivers, Konashen Indigenous District of Southern Guyana

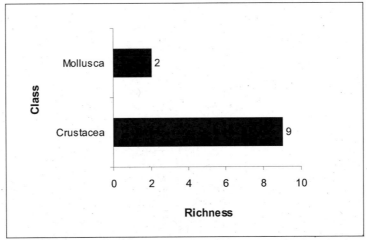

Figure 5.8. Speciess richness for classes collected during the 2006 RAP survey of the Acarai Mountains, Sipu, Kamoa and Essequibo rivers, Konashen Indigenous District of Southern Guyana.

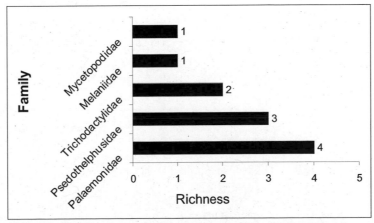

Figure 5.9. Species richness for families of aquatic macroinvertebrates collected during the 2006 RAP survey of the Acarai Mountains, Sipu, Kamoa and Essequibo rivers, Konashen Indigenous District of Southern Guyana.

Table 5.4. Presence-absence matrix of crustacean and mollusk species collected during the 2006 RAP survey in the Acarai Mountains, Sipu, Samoa and Essequibo rivers, Konashen Indigenous District of Southern Guyana. * Species observed but not collected.

Taxa	SR-01	SR-02	SR-03	SR-04	SR-05	AM-06	AM-07A	AM-07B	SR-08	KR-09	KR-10	KR-11	KR-12	WL-13	PS-14a	PS-14b	PF-15	MAKR-16	Total
CRUSTACEA																			0
Psedothelphusidae (3)																			0
Pseudothelphusidae sp. 1							1			1						1			3
Kingsleya cf. latifrons																1		1	2
Microthelphusa sp. 1												1							1
Trichodactylidae (2)																			0
Sylviocarcinus pictus										1	1	1	1						4
Valdivia serrata	1			1					1	1	1	1	1	1					8
Palaemonidae (4)																			0
Palaemonidae sp. 1												1							1
Euryrhynchus wrzeniowskii								1						1					2
Macrobrachium brasiliensis	1					1													2
Macrobrachium cf. nattereri	1	1	1	1				1	1	1	1	1	1	1			1	1	13
GASTROPODA																			0
Melaniidae (1)																			0
Doryssa sp. 1						1													1
BIVALVIA																			
Mycetopodidae (1)																			
Anodontites sp. 1																	1	1	
Total	3	1	1	2	0	2	1	2	2	4	3	5	3	3	0	2	1	2	37

Nearly half of the fish species we recorded are considered important subsistence fish resources, 20% are of sport fishing interest, and about 75% have high ornamental value.

Crustaceans and Mollusks (Gastropoda and Bivalvia)

Ten species grouped into three classes (Crustacea, Gastropoda, and Bivalvia) were recorded, of which Crustacea was the richest, with three families represented in the samples. Of these, Pseudothelphusidae exhibited the highest richness with four species, followed by Palaemonidae and Trichodactylidae with two species each. The mollusks were represented by only two species, a snail (*Doryssa* sp.), and a mussel (*Anodontites* sp.) (Figures 5.8 and 5.9).

The highest richness was concentrated in Focal Areas 2, 3 and 5, with five, six and four species respectively. Focal Areas 1 and 4 exhibited three species each (Table 5.4). The cluster analysis based on the Simpson Index of similarity identified Focal Area 5 as the locality most dissimilar, while Focal Areas 1 and 4 were the most related with a similarity of 0.67. Focal Areas 2 and 3 exhibited intermediate similarities, but closest to the group formed by Focal Areas 1 and 4, with a similarity index of about 0.5, which can be considered moderate (Figure 5.10).

The species accumulation curve exhibited sustained growth, starting with one species on day one, with no increase in the number of species on day two; from day three it increased on average by one species per day, until day nine, when the curve had still not stabilized completely, indicating that some species are yet to be recorded (Figure 5.11).

CONSERVATION RECOMMENDATIONS BY SITE

As previously indicated, all of the focal areas we sampled were in pristine, well preserved condition, probably as a result of being inside the Konashen COCA. The Acarai creek was the furthest and most inaccessible and therefore the best conserved. Although its diversity is not very high in comparison with other creeks in the Lower Essequibo, its conservation is very important given that it harbors unique species. It is important to notice that although the wealth of species in this creek seems low, its species numbers correspond to expected numbers in other streams located at similar elevations in the Guianas. Some of the Wai-Wai community members mentioned there was illegal mining in the region a few years back, but it currently appears to be a latent threat.

The Sipu and Kamoa rivers are very well conserved. The presence of numerous trunks, branches and trees crossed in the main channel are a clear indication of the low human disturbance and constitute an excellent refuge for fish. The zone most utilized by the Wai-Wai lies along the Essequibo River in the waters just above and below the Masakenari Village. Alexander et al. (2005) determined that four fishing waters below Masakenari, Amaci Falls, Kanaperu, Wanakoko and Mekereku, are of significant importance to the Wai-Wai community. In these areas, 26 species utilized by the community had previously been identified; we increased that number to 50 with the results of this RAP survey. All of these species are both of dietary and scientific (endemism, restricted distributions) interest, have elevated abundance, and show no evidence of overexploitation. The species faced with greatest subsistence fishing pressure are the aimara (*Hoplias macrophthalmus*) followed by the tiger fish (*Pseudoplatystoma fasciatum*). This fishing pressure could become problematic if fishing continues repeatedly at the same site. However, the Wai-Wai have well established fishing seasons and subsistence practices (hook and line) that are not as extractive as if they were to use gillnets or other, more deleterious fishing methods. The aimara is more obviously scarce closer to Masakenari Village, but populations are common both upstream and downstream from the village. We frequently observed the adults and the pre-adults in the main channel of the Essequibo, while the juveniles were more common in creeks. We collected little informa-

tion about the tiger fish (*Pseudoplatystoma fasciatum*), as it is a rather cryptic species with nocturnal or crepuscular habits, but we assume their status to be similar to the aimara. During the dry season the Wai-Wai use a natural ichthyiocide, hiari, to capture fish in the creeks and pocket waters of the Essequibo River – however this does not appear to constitute a threat because it has been done for so long in a sustainable manner.

General Conservation Recommendations

- The lower section of the Essequibo River, from Masakenari to the Amaci Falls, is of great diversity and use to the Wai-Wai, and remains to be sampled. For this reason, it is fundamental to conduct a second sampling expedition in the low water season (November-December) on the Wai-Wai fishing grounds which include, but are not limited to Amaci Falls, Kanaperu, Mekereku and Wanakoko. This would result in a more comprehensive and accurate species list, particularly in regard to the smaller-sized species.

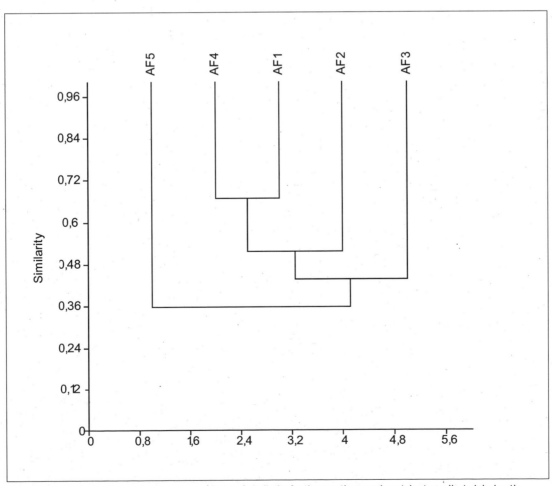

Figure 5.10. Cluster analysis using the Simpson Index of similarity for the aquatic macroinvertebrates collected during the 2006 RAP survey of the Acarai Mountains, Sipu, Kamoa and Essequibo rivers, Konashen Indigenous District of Southern Guyana.

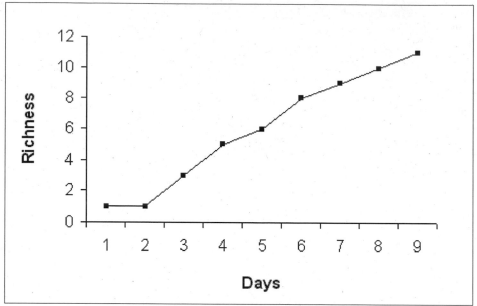

Figure 5.11. Accumulation curve of species of aquatic macroinvertebrates collected during the 2006 RAP survey of the Acarai Mountains, Sipu, Kamoa and Essequibo rivers, Konashen Indigenous District of Southern Guyana.

- Among the fish species we identified, there exists a considerable potential for aquarium and ornamental trade. However, to develop a plan that is sustainable and effective would require additional information on the present species' distribution and abundance. Taking this into account, it is recommended to complete an inventory of the fish species, and subsequently continue biological, ecological and market studies of these species.

- Begin biological, ecological and cultivation studies of the species that are important subsistence fishing resources. Particular focus should be given to aimara (*H. macrophthalmus*), tiger fish (*P. fasciatum*), kururú (*Curimata cyprinoides*) and the pakuchí or catabact pacú (*Myleus rhomboidalis*), among others.

- Design and implement a sustainable management plan, using the data from the studies outlined in the previous recommendations, which focuses on the Wai-Wai community's aquatic resources.

- Continue training parabiologists in the study, conservation and management of aquatic resources.

REFERENCES

Alexander, E., V. Antone, H. James. E. Joseph, A. Shushu, B. Suse, E. Mauruwanaru, R. Yamochi, C. Yukuma and R. Shoni. 2005. Preliminary results of a community-based fish mini-survey in the Konashen Indigenous District of Southern Guyana. Conservation International Guyana, Georgetown.

Conservation International (CI). 2003. Guayana Shield Conservation Priorities. Consensus 2002. Huber, O. and M. Foster (eds). Conservation International, Washington, DC.

Eigenamnn, C. 1912. The freshwater fishes of British Guiana, including a study of the ecological groupings of the species and the relation of the fauna of the plateau to that of the lowlands. Memories Carnegie Museum 5: 1-578.

Hammer., Ø., D.A.T. Harper and D. Ryan. 2001. Past. Paleontological statistics software package for education and data analysis. Palaeontologia Electronica 4 (1): 2 – 10.

Hardman, M., L.M. Page, M.H. Sabaj, J.W. Armbruster and J.H. Knouft. 2002. A comparison of fish surveys made in 1908 and 1998 of the Potaro, Essequibo, Demerara, and coastal river drainages of Guyana. Ichthyological Exploration of Freshwaters 13 (3): 225-238.

Lasso, C.A. 2002. Biological Diversity of Guianan Fresh and Brackish Water Fishes: Eastern Colombia, Venezuela, Guyana, Suriname, French Guiana and Northern Brazil. Conservation International-IUCN Netherlands Committee-Guiana Shield Initiative, Paramaribo.

Lasso, C.A., B. Chernoff and C. Magalhaes. 2003. Fishes and Freshwater Ecology. In: Huber, O. and M. Foster (eds.). Conservation Priorities for the Guayana Shield. 2002 Consensus. Conservation International, Washington, DC.

Mol. J.H. 2002. A Preliminary Assessment of the Fish Fauna and Water Quality of the Eastern Kanuku Mountains: Lower Kwitaro River and Rewa River al Corona Falls. Pp. 38-42. In: Montambault, J. R. and O. Missa (eds.). A Biodiversity Assessment of the Eastern Kanuku Mountains, Lower Kwitaro River, Guyana. RAP Bulletin of Biological Assessment 26. Conservation International, Washington, DC.

Watkins, G., W.G. Saul, C. Watson and D. Arjoon. 1997. Ichthyofauna of the Iwokrama Forest. Website: http://www.iwokrama.org.

Chapter 6

Amphibians and reptiles of the Acarai Mountains, and Sipu, Kamoa and Essequibo rivers in the Konashen COCA, Guyana

J. Celsa Señaris, Carlos A. Lasso, Gilson Rivas, Michelle Kalamandeen, and Elisha Marawanaru

SUMMARY

The herpetofauna recorded during the 2006 RAP survey of the Konashen Community Owned Conservation Area (COCA) in Guyana included 26 species of amphibians and 34 species of reptiles. The amphibians include representatives of the orders Gymnophiona (caecilians) and Anura (toads and frogs). More than half of the recorded anurans were treefrogs (Hylidae), with 13 species (54% of all recorded species), followed by the Leptodactylidae, with five species. Within reptiles, two species of crocodilians, three turtles, 14 lizards and 16 snakes were recorded. The blind snake *Typhlophis ayarzaguenai* represents the first record of this species for Guyana. The aquatic lizard *Neusticurus* cf. *rudis*, the snake *Helicops* sp., and the caecilian may also represent new records for the Guyana herpetofauna, but require additional taxonomic reviews. The three focal areas explored during this survey differed significantly in their faunistic composition. The surveyed region appears intact and in pristine condition, particularly the Acarai Mountains and the flooded forests adjacent to the main channels of the Kamoa and Sipu rivers. The area of the Essequibo River closest to Masakenari and Akuthopono villages showed a lower abundance of medium-to-large bodied reptiles, turtles and caimans, which are a part of the Wai-Wai diet, but populations of other reptiles and amphibians seemed to be in good condition. Taxa used by local communities should be monitored for signs of overexploitation.

INTRODUCTION

The knowledge of Guayana Shield herpetofauna, while fragmentary, is increasing rapidly, particularly with respect to the highlands or Pantepui, over 150 m asl (McDiarmid and Donnelly 2005, Avila-Pires 2005). Señaris and Avila-Pires (2003) list only 40 localities with a medium or high degree of herpetological exploration, including four in Guyana (Raleigh Falls, Kabalebo, Iwokrama, and Bartica), whereas the knowledge of the herpetofauna of most of the low- and medium-elevation lands in the Guayana Shield remains poor or none. During the last year several herpetological surveys at different sites in Guyana have taken place, demonstrating rich biodiversity and high levels of endemism, associated mainly with upper elevations and highlands (e. g. Cole and Kok 2006, Donnelly et al. 2005, Ernst et al. 2005, Kok et al. 2006, MacCulloch and Lathrop 2002, 2005; MacCulloch et al. 2006).

Despite the increasing knowledge of amphibians and reptiles in Guyana, the southern part of the country has yet to be explored. Southern Guyana has been cited as a high research priority because it harbors large, contiguous forests, and a high diversity of habitats. In particular, the herpetofauna of the Acarai Mountains and the upper Essequibo River have never been surveyed. Theses areas have potentially high species richness, indicated by the presence of both Guayana Shield and Amazonian faunistic elements. In an effort to increase the knowledge of this area, during a Rapid Assessment survey (RAP) in October 2006, we collected and observed herpetofauna of the Konashen COCA, and the results of this survey are presented here.

METHODS

We surveyed amphibians and reptiles during the period of October 15-26, 2006. During the first week of the survey observations were made only by MK and EM. The first step of our work included a preliminary survey of the study sites in an attempt to identify the primary habitats and microhabitats associated with water systems – rivers, streams, lagoons – and prioritize the survey activities within the short sampling time (Scott 1994). We used a combination of opportunistic surveys and "Visual Encounter Survey" (VES) (Crump and Scott 1994, Doan 2003), both during the day and night, using the main course of the river as our transect, in addition to random, long walks in the aquatic/terrestrial transition zone (margins of bodies of water), and long walks between different study sites. The herpetological sampling was restricted largely to the main channels of rivers and their tributaries, and to the adjacent vegetation, with the exception of occasional collections made by other RAP team members (principally the insect team). The length of transects varied depending on the characteristics of each site and the logistics and, as a result, the sampling effort between sites was not equal (Table 6.1). The opportunistic surveys and the occasional collections made by other RAP team members have not been taken into account in the estimation of sampling effort.

Adults and juvenile amphibians and reptiles were captured manually once visually located. Tadpoles were collected using fine mesh nets, or were opportunistically collected by the fish team. For each specimen, we assigned a field number and noted the locality and date of collection, preliminary identification, general description of the habitat or microhabitat, and coloration in life. Some of the specimens were photographed live by Piotr Naskrecki, and the herpetological team kept some of the photographs as records of species collected and/or observed. The samples were anesthetized and fixed using 10% formol, and preserved in 70% ethanol. The majority of collected specimens have been deposited at the Center for the Study of Biological Diversity, University of Guyana, Georgetown, and a smaller reference collection has been deposited in the Museo de Historia Natural La Salle (MHNLS), Caracas, Venezuela for final identification. In addition, we conducted non-structured interviews with local field guides who accompanied us. This resulted in additional records of the herpetofauna, particularly for the medium-to-large reptiles. The amphibian list and taxonomy follows the recent changes proposed by Faivovich et al. (2005), Frost et al. (2006), and Grant et al. (2006).

RESULTS

General Results

We recorded 26 species of amphibians and 34 species of reptiles for the entire study area (all three focal areas). Most amphibians belonged to the order Anura (with 25 species of frogs and toads), and we collected only one species of

Table 6.1. Herpetological sampling schedule during the RAP survey of the Konashen COCA in Guyana. OS = opportunistic survey; VES = Visual Encounter Survey.

DATE (2006)	Focal Area	Locality	Day time	Habitat/Effort
10/14		Essequibo Mazakerani - Sipu River	diurnal	OS: Principal channel by boat - gallery forest
10/15	Sipu River - base Acarai Mountains	Sipu River camp - Acarai Mountains	diurnal	VES: Forest (3:10 hours)
			nocturnal	VES: Forest adjacent to Sipu camp (2:43 hours)
10/16		Sipu River	diurnal	VES: Small creek (3:20 hours)
			nocturnal	VES: Forest and isolated pool in forest (3:10 hours)
10/17		Sipu River camp - Acarai Mountains	diurnal	VES: Forest (3:20 hours)
		Acarai site	nocturnal	VES: Acarai creek (2:20 hours)
10/18		Acarai site	diurnal	VES: Acarai creek (2:30 hours)
		Acarai site	nocturnal	VES: Forest (1:15 hours)
10/19		Acarai site-Sipu River	diurnal	VES: Forest (2:50 hours)
			nocturnal	VES: Small creek-Sipu River (2:10 hours)
10/20		Sipu River - Essequibo	diurnal	OS: Principal channel by boat - gallery forest
10/21	Kamoa River	Essequibo - Kamoa River	diurnal	OS: Principal channel by boat - gallery forest
		Kamoa River	nocturnal	OS: Principal channel by boat - gallery forest
10/22		Kamoa camp	diurnal	VES: Forest and small creeks (3.50 hours)
		Kamoa River	nocturnal	VES: Gallery forest (2:40 hours)
10/23		Kamoa River - Mazakenari	diurnal	OS: Principal channel by boat - gallery forest
10/23	Essequibo at Akuthopono	Akuthopono	nocturnal	OS: around village
10/24		Akuthopono	diurnal	VES: around village and forest (3:20 hours)
10/25		Wanakoko Lake-Essequibo River	nocturnal	OS: Principal channel by boat - gallery and flooded forest
10/26		Essequibo River - Akuthopono forest	diurnal	VES: Forest (2:30 hours)
		Essequibo River - Akuthopono forest	nocturnal	VES: Forest (1:50 hours).

Gymnophiona. More than a half of the anurans were tree-frogs (Hylidae), with 13 species (54% of the total), followed by the Leptodactylidae (five species), toads (Bufonidae, three species), poison arrow frogs (Dendrobatidae, two species), and single representatives of the families Centrolenidae and Pipidae (Table 6.2). Within reptiles, we recorded two species of crocodilians, three turtles, 14 lizards, and 16 snakes. The lizards belonged to seven families, and the snakes were dominated by colubrids (Table 6.3). All large reptiles recorded, the two species of crocodiles and the three turtles, are a part of the Wai-Wai diet.

The sampling stations in three focal areas explored during this survey show significant differences in their faunistic composition (Table 6.4, Figures 6.1 and 6.2), and are discussed below.

RESULTS FOR EACH FOCAL AREA

Focal area 1: Sipu River - Acarai Mountains

This focal area was situated between the Sipu River and the base of the Acarai Mountains (250-270 m a.s.l.), and was characterized by sandy, oligotrophic soils, with lowland evergreen, deciduous forest, with no signs of seasonal inundation. Of the three focal areas we surveyed during this RAP, this site had the highest species richness, with 19 species of amphibians and 29 species of reptiles (Table 6.4). In addition, the abundance observed for some species, both of amphibians and reptiles, was remarkably high compared to other areas surveyed during this study. Thirty-eight percent of all amphibians and reptiles recorded during this RAP survey were

found only in the Sipu River-Acarai Mountain focal area, and many of them seemed to be restricted to this locality. These results reflect the area's pristine condition, where some habitat types — non-flooded forests, small rocky streams, forest ponds — can be found only in this focal area.

The hylid treefrogs were the richest group we observed in the Sipu River-Acarai Mountains, with ten recorded species, followed by terrestrial frogs (Leptodactylidae), with four species of the genus *Leptodactylus*. The poison arrow frogs (Dendrobatidae) were represented by *Ameerega picta* and *Dendrobates tinctorius*. The caecilian was found only in the non-flooded forest of the Acarai Mountains. Another interesting group of amphibians found in this focal area included the monkey frogs of the genus *Phyllomedusa,* with two collected species, and an additional species, *P. bicolor,* observed but not collected by MK (Table 6.2).

Sixty-eight percent of all the reptiles recorded during this RAP survey were found in the Sipu River-Acarai Mountain focal area, and 11 (38%) of them were exclusive to this site. In addition to this richness, the relative abundance of certain species found in this site was higher than in the other focal areas. This was especially evident in small-bodied reptile species, where we observed more than 20 individuals in only 30 minutes of sampling effort (e.g., the diving lizard *Uranoscodon superciliosus* in the Sipu River, or the streamside lizards *Neusticurus* cf. *rudis* in the rocky streams in Acarai). We also recorded between 6-15 dwarf caimans *Paleosuchus trigonatus* in 40-45 minutes of the nocturnal survey on the Sipu River. The black caiman *Melanosuchus niger,* was seen only in the main channel of the Sipu River (Table 6.3).

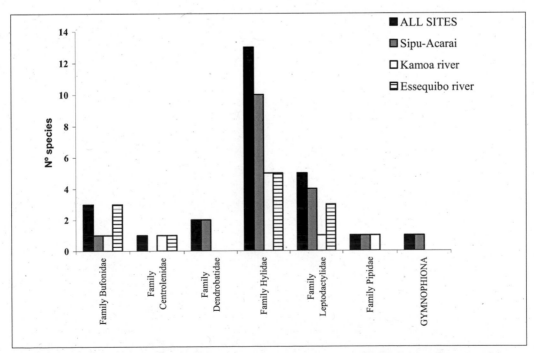

Figure 6.1. Number of amphibian species, by family, recorded at each site surveyed during the 2006 RAP survey of the Konashen COCA, Guyana.

Table 6.2. Amphibians recorded during the October 2006 RAP survey of the Konashen COCA, Guyana.

TAXA	FOCAL AREA		
	Sipu-Acarai Mountains	Kamoa River	Essequibo River
ORDER ANURA			
Family Bufonidae			
Chaunus marinus (Linnaeus, 1758)			x
Rhaebo guttatus Schneider, 1799			x
Rhinella margaritifera complex (Laurenti, 1758)	x	x	x
Family Centrolenidae			
Allophryne ruthveni Gaige 1926		x	x
Family Dendrobatidae			
Ameerega picta (Tschudi, 1838)	x		
Dendrobates tinctorius (Cuvier, 1797)	x		
Family Hylidae			
Hypsiboas boans (Linnaeus,1758)	x	x	x
Hypsiboas cinerascens (Spix, 1824)	x		
Hypsiboas calcaratus (Troschel,1848)	x	x	x
Hypsiboas geographicus (Spix,1824)		x	x
Hypsiboas ornatissimus (Noble,1923)	x		
Hypsiboas wavrini (Parker,1936)		x	
Osteocephalus cabrerai Cochran et Goin, 1970		x	
Osteocephalus cf. *leprieurii* (Duméril et Bribon,1841)	x		
Osteocephalus sp. 1	x		x
Phyllomedusa bicolor (Boddaert, 1772)	x		
Phyllomedusa hypocondrialis (Daudin, 1800)	x		
Phyllomedusa vaillanti Boulenger, 1882	x		
Scinax ruber (Laurenti, 1768)	x		x
Family Leptodactylidae			
Leptodactylus knudseni Heyer, 1972	x	x	x
Leptodactylus mystaceus (Spix 1824)	x		
Leptodactylus rhodomystax Boulenger 1884	x		
Leptodactylus sp. 1 (*wagneri* group)	x		x
Leptodactylus sp. 2			x
Family Pipidae			
Pipa pipa (Linnaeus, 1758)	x	x	
ORDER GYMNOPHIONA			
Family Caecilidae			
Caecilidae sp.	x		
Total 26	19	9	12

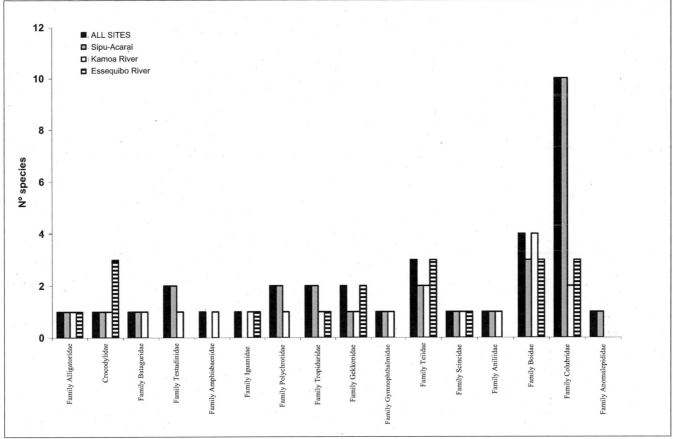

Figure 6.2. Number of reptile species, by family, recorded at each site during the 2006 RAP survey of the Konashen COCA, Guyana.

Focal area 2: Kamoa River

At this site, situated on the north bank of the Kamoa River at 250 m a.s.l. and annually inundated, we were able to conduct only two effective sampling days, augmented by opportunistic collecting and/or observations carried out by other RAP team members. Although this focal area had the lowest sampling efforts, we were able to record nine species of amphibians and 19 species of reptiles. The abundance of certain amphibians and reptiles at the Kamoa River was relatively high, and two species, the crested forest toads *Bufo margaritifera* complex and the chicken frog *Leptodactylus knudseni*, were unique to this area. The abundance of *Paleosuchus trigonatus* was similar to that observed from the Sipu River. We also recorded the emerald tree boa *Corallus caninus* and the worm lizard *Amphisbaena vanzolinii* (Table 6.3).

Focal area 3: Wanakoko Lake, Essequibo River and Akuthopono and Masakenari villages.

Because of the importance of these areas to the Wai-Wai community, we spent five days, our greatest sampling effort, at these localities. Nevertheless, we recorded the lowest species richness of reptiles, and only a moderate diversity of amphibians in this area (Table 6.4). In the villages of Akuthopono and Masakenari we observed a large abundance of the common cane toad *Chaunus marinus*, the smooth-sided toad *Rhaebo guttatus*, and the frog *Leptodactylus* sp. 2 as well as the black spotted skink *Mabuya nigropuntata*, the

bridled gecko *Gonatodes humeralis*, and turnip tailed gecko *Thecadactylus rapicauda* in the houses, and whiptail lizards *Ameiva ameiva* and *Kentropix calcarata* at sites nearby. These species are generally associated with, and abundant at, habitats that are impacted by human activity.

In the principal channel of the Essequibo River between Akuthopono village and Wanakoko Lake we observed only one individual of the dwarf caiman, which is indicative of a significant use of this species by the community, and a notable decrease in the density of its local population.

DISCUSSION

The results of this short, dry season survey at the base of the Acarai Mountains and the Sipu, Kamoa and Essequibo rivers undoubtedly represent only a fraction of the herpetofauna of this area, and more work should be done to discover the real richness of its amphibians and reptiles. Although the sampling efforts were different between the three focal areas explored, we think that the Acarai Mountains – Sipu River focal area is the herpetologically richest site of the Konashen COCA of southern Guyana. The elevation ranges in the Acarai Mountains, and its unique habitats – non-flooded forest, mountains streams, etc. – probably harbor a number of endemic and undescribed species, making this area extremely important for future herpetological research.

Table 6.3. Reptiles recorded during the October 2006 RAP survey of the Konashen COCA, Guyana.

TAXA	FOCAL AREA		
	Sipu-Acarai Mountains	Kamoa River	Essequibo River
ORDER CROCODYLIA			
Family Alligatoridae			
Paleosuchus trigonatus (Schneider, 1801)	x	x	x
Family Crocodylidae			
Melanosuchus niger (Spix 1825)	x		
ORDER TESTUDINES			
Family Bataguridae			
Rhinoclemmys punctularia (Daudin, 1802)	x	x	
Family Testudinidae			
Chelonoidis carbonaria (Spix, 1824)	x	x	
Chelonoidis denticulata (Linnaeus, 1766)	x		
ORDER SQUAMATA			
Family Amphisbaenidae			
Amphisbaena vanzolinii Gans 1963		x	
Family Iguanidae			
Iguana iguana Linnaeus, 1758		x	x
Family Polychrotidae			
Anolis punctatus Daudin 1802	x		
Norops chrysolepis Troeschel, 1845	x	x	
Family Tropiduridae			
Plica plica (Linnaeus, 1758)	x		
Uranoscodon superciliosus (Linnaeus, 1758)	x	x	x
Family Gekkonidae			
Gonatodes humeralis (Guichenot, 1855)	x	x	x
Thecadactylus rapicauda (Houttuyn, 1782)			x
Family Gymnophthalmidae			
Neusticurus cf. *rudis* Boulenger 1900	x	x	
Family Teiidae			
Ameiva ameiva (Linnaeus, 1758)	x	x	x
Kentropyx calcarata Spix, 1825	x	x	x
Tupinambis teguixin (Linnaeus, 1758)			
Family Scincidae			
Mabuya nigropunctata Spix, 1825	x	x	x
Family Aniliidae			
Anilius scytale (Linnaeus, 1758)	x	x	
Family Boidae			
Boa constrictor Linnaeus, 1758	x	x	x
Corallus caninus (Linnaeus, 1758)		x	
Corallus hortulanus (Linnaeus, 1758)	x	x	x
Eunectes murinus Linnaeus, 1758	x	x	x
Family Colubridae			
Chironius scurrulus (Wagler, 1824)	x	x	
Atractus torquatus (Duméril, Bibron y Duméril, 1854)	x		x
Dipsas indica Laurenti, 1768	x		
Erythrolamprus aesculapii (Linnaeus, 1766)	x		
Helicops angulatus (Linnaeus, 1758)	x	x	x
Helicops sp.	x		
Imantodes cenchoa (Linnaeus, 1758)	x		
Siphlophis compressus (Daudin 1803)	x		x
Taeniophalus brevirostris (Peters 1863)	x		
Leptodeira annulata (Hallowell, 1845)	x		
Family Anomalepididae			
Typhlophis ayarzaguenai Señaris, 1998	x		
Total 34	29	19	14

Table 6.4. Richness of amphibians and reptiles in the three focal areas of the 2006 RAP survey in the Konashen COCA, Guyana. In parentheses are the numbers of species exclusive to each area.

TAXA	FOCAL AREA		
	Sipu-Acarai Mountains	Kamoa River	Essequibo River
Amphibians	19 (11)	9 (3)	12 (3)
Reptiles	29 (11)	19 (2)	14 (1)

The entire region we sampled is virtually intact and in pristine condition, particularly the Acarai Mountains and the flooded forests adjacent to the main channels of the Kamoa and Sipu rivers. The area of the Essequibo River closest to Masakenari and Akuthopono villages showed a reduction in the abundance of medium-to-large bodied reptiles, turtles and caimans, which are part of the Wai-Wai diet, but other reptile and amphibian communities appeared to be in good condition.

The results from the Konashen COCA contribute to the knowledge of the herpetofauna of the upper Essequibo River and Guyana, and several species found there are of special taxonomic, ecological, and conservation interest. The blind snake *Typhlophis ayarzaguenai* represents the first record for the country; the aquatic lizard *Neusticurus* cf. *rudis*, the snake *Helicops* sp. and the caecilian still require a comprehensive taxonomic evaluation, and may represent new additions to the herpetofauna of Guyana (the caecilian is likely to be a new, undescribed species). *Amphisbaena vanzolinii* is a little-known species of high research interest, previously known only from a few specimens from Guyana (type locality Marudi), Suriname, and adjacent areas of Brazil.

Other recorded taxa are of particular conservation interest. Recently Wollenberg et al. (2006), supported by Grant et al. (2006), synonymized *Dendrobates azureus* (listed as Vulnerable by the IUCN Red List, and included in Appendix II of CITES) with *D. tinctorius* (Least Concern species). The tortoises of the genus *Chelonoidis* are included in Appendix II of CITES, and *C. carbonaria* is listed as Vulnerable by the IUCN Red List (IUCN 2007). The black caiman *Melanosuchus niger* is classified as Low Risk, but its conservation is recommended, and it is included in Appendix I of CITES. The dwarf caiman *Paleosuchus trigonatus* and the emerald tree boa *Corallus caninus* are included in Appendix II of CITES.

CONSERVATION RECOMMENDATIONS

Based on the observations obtained during the 2006 RAP survey of the Konashen COCA in Guyana we make the following recommendations:

1. The results of this survey are preliminary and we suspect that a much greater diversity of amphibians and reptiles is to be found here. For these reasons, we recommend more extensive sampling of the entire region, including sampling during both the rainy and dry seasons. Also, particular attention should be given to the Acarai Mountains where we expect a high species richness and a possible center of endemism of amphibians and small reptiles.

2. We recommend specific studies of the use of large reptiles (e.g., black caimans and tortoises) by the local human population. The abundance of medium-to-large bodied reptiles that are hunted by the Wai-Wai community should be monitored, and a sustainable management plan must by developed to guarantee local conservation of these species.

3. Many of the amphibians and reptiles recorded during this survey are of great eco-tourism potential and/or are important in the pet trade. We recommend considering these taxa in future ecotourism plans. We also recommend additional biological studies of amphibians and reptiles that are especially important for these activities.

4. Continue and intensify the training of members of the local Wai-Wai community in the study, management, conservation and monitoring of the local herpetofauna.

REFERENCES

Avila-Pires, T.C.S. 2005. Reptiles. *In*: Hollowell, T. and R. Reynolds (eds). Checklist of the Terrestrial vertebrates of the Guiana Shield. Bulletin of the Biological Society of Washington 13: 25-40.

Cole, C. and P. Kok. 2006. A New Species of Gekkonid (Sphaerodactylinae: *Gonatodes*) from South America. American Museum Novitates 3524: 1-16.

Crump, M.L. and N.J. Scott. 1994. Visual Encounnter Surveys. pp. 84-92. *In*: Heyer, W.R., M.A. Donnelly, R.W. McDiarmid, L.C. Hayerk, and M.S. Foster (eds.). Measuring and monitoring Biological Diversity. Standard Methods for Amphibians. Smithsonian Institution Press, Washington.

Doan, T.M. 2003. Which methods are most effective for surveying rain forest herpetofauna? Journal of Herpetology 37(1): 72-81.

Donnelly, M.A., M.H. Chen and G.G. Watkins. 2005. The Iwokrama Herpetofauna: An Exploration of Diversity in a Guyanan Rainforest. pp. 428-460. *In*: Donnelly, M.A., B.I. Crother, C. Guyer, M.H. Wake and M.E. White (eds.). Ecology & Evolution in the Tropics: A Herpetogical Perspective. The University of Chicago Press, Chicago.

Ernst, R., M-O. Rödel and D. Arjoon. 2005. On the cutting edge – The anuran fauna of the Mabura Hill Forest Reserve, Central Guyana. Salamandra 41(4): 179-194.

Faivovich, J., C.F.B. Haddad, P.C.A. García, D.R. Frost, J.A. Campbell and W.C. Wheeler. 2005. Systematic review of the frog family Hylidae, with special reference to Hylinae: phylogenetic analysis and taxonomic revision. Bulletin of the American Museum of Natural History 294: 1-240.

Frost, D.R., T. Grant, J. Faivovich, R.H. Bain, A. Haas, C.F.B. Haddad, R.O. De Sá, A. Channing, M. Wilkinson, S.C. Donnellan, C.J. Raxworthy, J.A. Campbell, B.L. Blotto, P. Moler, R.C. Drewes, R.A. Nussbaum, J.D. Lynch, D.M. Green and W.C. Wheeler. 2006. The Amphibian Tree of Life. Bulletin of the American Museum of Natural History 297: 1- 370.

Grant, T., D.R. Frost, J.P. Caldwell, R. Gagliardo, C.F.B. Haddad, P.J.R. Kok, D.B. Means, B.P.Noonan, W.E. Schargel and W.C. Wheeler. 2006. Phylogenetic systematics of dart-poison frogs and their relatives (Amphibia: Athesphatanura: Dendrobatidae). Bulletin of the American Museum of Natural History 299: 1-262.

IUCN. 2007. IUCN Red List of Threatened Species. <www.iucnredlist.org>. Downloaded in December 2007.

Kok, P., H. Sambhu, I. Roopsind, G. Leglet and G. Bourn. 2006. A new species of *Colostethus* (Anura: Dendrobatidae) with maternal care from Kaieteur National Park, Guyana. Zootaxa 1238: 35-61.

MacCulloch, R. and A. Lathrop. 2002. Exceptional diversity of *Stefania* on Mount Ayanganna, Guyana: three new species and new distributional records. Herpetologica 58: 327-346.

MacCulloch, R. and A. Lathrop. 2005. Hylid frogs from Mount Ayanganna, Guyana: new species, redescriptions, and distributional records. Phyllomedusa 4(1): 17-37.

MacCulloch, R., A. Lathrop and S. Khan. 2006. Exceptional diversity of *Stefania* II: Six species from Mount Wokomung, Guyana. Phyllomedusa 5(1): 31-41.

McDiarmid, R.W. and M.A. Donnelly. 2005. The Herpetofauna of the Guyana Highlands: Amphibians and Reptiles of the Lost World. pp. 461-560. *In*: Donnelly, M.A., B.I. Crother, C. Guyer, M.H. Wake and M.E. White (eds.). Ecology and Evolution in the Tropics: A Herpetological Perspective. University of Chicago Press, Chicago.

Scott, N.J. 1994. Complete Species Inventories. pp. 78-84. *In*: Heyer, W.R., M.A. Donnelly, R.W. McDiarmid, L.C. Hayerk and M.S. Foster (eds.). Measuring and Monitoring Biological Diversity. Standard Methods for Amphibians. Smithsonian Institution Press, Washington.

Señaris, J.C. and T. Avila-Pires. 2003. Anfibios y reptiles. pp. 11-13. *In*: Huber, O., and M.N. Foster (eds). Prioridades de Conservación para el Escudo de Guayana, Consenso 2002. Conservación Internacional, Washington DC.

Wollenberg, K.C., M. Veith, B.P Noonan and S. Lötters. 2006. Polymorphism versus species richness. Systematics of large *Dendrobates* from the eastern Guiana Shield (Amphibia: Dendrobatidae). Copeia 2006 (4): 623-629.

Chapter 7

Birds of the Konashen COCA, Southern Guyana

Brian J. O'Shea

SUMMARY

Avifaunal surveys were conducted around two sites in the Konashen Community Owned Conservation Area (COCA) between 6 and 28 October 2006. The purpose of the surveys was to obtain a baseline estimate of the avian species richness of the area, and to provide information on the population status of several bird species important to the indigenous people inhabiting the region. Birds were surveyed on foot and by boat during all daylight hours of the study period. Cassette recorders and directional microphones were used to document the avifauna; several species were also documented with a video camera. Species richness was high at both sites; a combined total of 319 species was tallied over the study period. Documentation was obtained for the majority of species encountered. The avifauna was largely composed of species that would be expected to occur in a Guianan lowland forest, and included 27 species that are endemic to the Guayana Shield. There was a high degree of habitat heterogeneity within each site. Six distinct habitats were identified, only two of which were shared between the two study sites. As a result, the avian diversity was higher than expected for the size of the area surveyed. It is probable that at least 400 bird species, or more than half of the number known to occur in Guyana, may be found in the Konashen COCA.

The survey recorded Large-headed Flatbill (*Ramphotrigon megacephalum*), a new record for Guyana and a range extension of approximately 900 km. Populations of parrots, guans, and curassows, all of which are important to the Wai-Wai inhabitants of the region and are of global conservation concern, seemed healthy. Fourteen species of parrots were observed, including Scarlet Macaw (*Ara macao*), a CITES Appendix 1 species, and Blue-cheeked Parrot (*Amazona dufresniana*), listed as Near Threatened (IUCN 2006). Some of the larger parrot species are hunted by local people, but the effects of this hunting appear to be negligible. There was no evidence that parrots in the area are subjected to the intense trapping pressure that exists in more accessible regions of the Guayana Shield. This impression was corroborated through interviews with the Wai-Wai. Spix's Guan (*Penelope jacquacu*) and Black Curassow (*Crax alector*) were common at both survey sites, suggesting that their regional populations are not threatened by current levels of hunting pressure from the local community.

The remarkable avian diversity of the Konashen COCA does not seem to be faced with any immediate threats. The vast majority of bird species in the area are also found in the surrounding region and beyond, and their global populations are not threatened. Parrots and large game birds, though not currently threatened at a regional level, are of global conservation concern. Care should be taken to forestall local declines in their populations. Monitoring is not recommended at the present time, since these species are not amenable to standardized survey methods. Instead, the Wai-Wai community should continue to avoid trapping parrots for the pet trade, and should deny trappers entry to the Konashen COCA. Rather than monitor populations of large game birds, the community should establish a rotation system to distribute the effects of subsistence hunting over as large an area as possible. This should involve the closing of most of the Konashen COCA to hunting at any given time. Finally, the Wai-Wai should aggressively exclude illegal Brazilian miners from their territory, and (if necessary) seek assistance from the government of Guyana to maintain sovereignty over their land.

INTRODUCTION

Bird communities are generally reflective of environmental conditions. In lowland tropical forest, many of the larger species are important seed dispersers and predators, and thus have a substantial effect on forest dynamics. They are also important food sources for other animals and people. Healthy populations of these larger species are indicative of a relatively intact, undisturbed ecosystem. Since many of these species are conspicuous when they are common, it is comparatively easy to assess their population status, even within the constraints of a rapid inventory. For the purposes of such inventories, birds are excellent indicators – they are primarily diurnal and can therefore be surveyed easily, they are generally easy to detect and identify, and the richness of bird communities tends to correlate positively with other measures of biodiversity.

In contrast to many other taxonomic groups, the avifauna of Guyana is well known (Braun et al. 2000). Numerous inventories have been conducted at multiple sites in the country, and a picture of avian distributions across Guyana is beginning to emerge as the results of those studies are published (e.g., Braun et al. 2003; Finch et al. 2002; Robbins et al. 2004, 2007; Ridgely et al. 2005; O'Shea et al. 2007). Most of the interior of Guyana is still covered by unbroken tropical moist forest and is sparsely populated. Accordingly, the avifauna is rich in species, and previous surveys have found that many sites support healthy populations of species that are of global conservation concern, such as large raptors, cracids, and parrots.

Although current levels of human pressure on Guyana's natural wealth are rather low, the need to identify areas of exceptional biodiversity within the country becomes ever more urgent as Guyana's infrastructure develops. Few formally protected areas currently exist in the country. Previous survey work in the Kanuku Mountains (Parker et al. 1993, Finch et al. 2002) has led to recommendations for protected status in that area, but protective measures have yet to be implemented. The Konashen COCA is one of the most remote regions of Guyana; adjoining areas of Brazil are similarly isolated from the infrastructure of that country, ensuring that the Konashen COCA faces no immediate threats. However, the ongoing construction of a highway across northern Brazil poses a potential threat to the Konashen COCA in the near future. Illegal miners from Brazil are a persistent presence throughout the interior regions of the Guianas, a situation that can be expected to worsen as the highway advances. Documentation of the biodiversity of the Konashen COCA is thus a timely endeavor.

The avifauna of the Acarai Mountains was surveyed in 1999-2000 by researchers from the Smithsonian Institution and University of Kansas (Robbins et al. 2007). Naka et al. (2006), in an exhaustive account of the avifauna of Roraima, Brazil, mentioned several survey localities that are quite close to the Konashen COCA, but their paper does not treat the avifauna of those localities in detail.

We surveyed the avifauna at two localities within the Konashen COCA between 6 and 28 October 2006:

Site 1. Acarai Mountains N 01° 23' 12.5" W 058° 56' 46.0" elevation 250 m, 6-10 and 15-18 October. This locality includes the satellite camp, New Romeo's Camp (N 01° 21' 19.0" W 058° 57' 25.5" elevation 526 m), surveyed between 11-15 October.

Site 2. Kamoa River N 01° 31' 51.8" W 058° 49' 42.4" elevation 240 m, 19-28 October.

Birds were surveyed by boat and on foot during all daylight hours of the study period. Throughout the study period, we attempted to identify and survey as many different habitats as possible, devoting equal effort to each habitat type as logistics allowed. Our survey coincided with the long dry season, but rainfall was nevertheless substantial, particularly during the first half of the study period.

A complete list of the birds encountered in the Konashen COCA appears in Appendix 3.

METHODS

At all sites, birds were observed during all daylight hours. Observation methods consisted of walking along trails to locate and identify birds. Coverage of the trail system at each site was intended to maximize observation time in each habitat type. Typically, one observer (BJO) would leave camp 30-60 minutes before first light, to be joined by Wai-Wai parabiologists 2-3 hours later. Morning excursions typically lasted until 10:00 – 11:00, by which time bird activity had decreased considerably. The field team also surveyed areas near the camps on most afternoons between approx. 15:30 and 18:00. Birds were observed opportunistically at all other times of the day.

At the Kamoa site, birds were also surveyed by boat on four mornings by floating down the river at dawn with the boat motor turned off.

Birds were documented using a Sony TCM-5000EV cassette recorder with a Sennheiser ME-66 shotgun microphone and a Sony TC-D5 Pro-II stereo cassette recorder with a Sennheiser ME-62 omni-directional microphone and Telinga parabolic reflector. Due to the pronounced decrease in bird vocal activity after mid-morning, the majority of recordings were made between one hour before and three hours after sunrise on each day. Tape recordings are deposited at the Macaulay Library, Cornell Lab of Ornithology.

At the Kamoa site, several species were recorded opportunistically with a Sony ZR-500 Mini-DV video cassette recorder.

For each site, approximate numbers of each species were recorded on a daily basis, and relative abundances were determined from these data, as follow:

A: abundant; observed every day; always 20 or more individuals, pairs, or groups encountered daily in appropriate habitat

C: common; observed on at least 90% of days at each site; minimum of 5 individuals, pairs, or groups encountered daily in appropriate habitat

F: fairly common; observed on at least 50% of days at each site; average 1-5 individuals, pairs, or groups encountered daily in appropriate habitat

U: uncommon; observed on fewer than 50% of days at each site; average fewer than one individual, pair, or group encountered daily in appropriate habitat.

For the majority of species that were encountered at both sites, abundances between sites did not vary significantly; therefore, data from the two sites was pooled for the final species list. This list (Appendix 3) also includes information on habitat associations and elevation restrictions for each species. Species that were only observed during travel between survey sites were included in the list as well, but relative abundances were not calculated for these species.

RESULTS

We encountered a total of 319 species during the survey: 250 species were found at the Acarai site, and 232 species at Kamoa. This remarkable diversity was likely related to the high degree of habitat heterogeneity in the Konashen COCA – six distinct habitats were identified, only two of which were shared between the two survey sites. As a result, the avifauna of the two sites, while overlapping broadly, nevertheless contained distinct elements. Seventy-two species were observed only at the Acarai site, and 55 species only at the Kamoa site; the vast majority of such "unique" species were restricted to habitats that were only found at their respective sites (see Appendix 3). Overall, the avian species richness of the Konashen COCA is high. It is certain that well over 400 species, or more than half of the species known to occur in Guyana, occur in the area.

The avifauna of the Konashen COCA yielded few surprises, and consisted of a typical Amazonian/Guianan lowland tropical moist forest assemblage. The majority of species were relatively rare (fewer than five individuals, pairs, or groups encountered daily), a typical abundance pattern in undisturbed regions of lowland forest. At elevations above approx. 800 m at the Acarai site, we encountered a suite of species with highland affinities, such as *Megascops guatemalae* (Vermiculated Screech-Owl), *Aeronautes montivagus* (White-tipped Swift), *Colibri delphinae* (Brown Violet-ear), *Aulacorhynchus derbianus* (Chestnut-tipped Toucanet), and *Hylophilus sclateri* (Tepui Greenlet). We also found several more widespread species that were here restricted to forest above 800 m, such as *Dysithamnus mentalis* (Plain Antvireo), *Herpsilochmus rufomarginatus* (Rufous-winged Antwren), *Cyclarhis gujanensis* (Rufous-browed Peppershrike), and *Setophaga ruticilla* (American Redstart). None of the aforementioned species were observed at the Kamoa site. The majority of lowland species at the Acarai site were recorded up to the maximum elevation surveyed (1050 m). Species restricted to riverine forest were not encountered above approx. 300 m.

Many of the expected lowland forest species that we failed to encounter during the survey are generally rare and/or inconspicuous, and are unlikely to be observed over the course of a short survey with few observers during only one season. It is exceedingly likely that their presence would be revealed with continued survey effort.

Thirty-two species of Guayana Shield endemics were observed during the study period (Table 7.1). Many of those species are geographic representatives of widespread species complexes, and are widely distributed in interior lowland forests of the Guayana Shield.

The most noteworthy record of the study period was our observation of *Ramphotrigon megacephalum* (Large-headed Flatbill), which was seen and tape-recorded near the Acarai camp on October 8th. We found at least two pairs in a limited area of dense *Guadua* sp. bamboo along a creek

Table 7.1. Bird species (32) recorded during the RAP survey that are endemic to the Guayana Shield.

Black Curassow (*Crax alector*)
Caica Parrot (*Gypopsitta caica*)
Blue-cheeked Parrot (*Amazona dufresniana*)
Rufous-winged Ground-Cuckoo (*Neomorphus rufipennis*)
Guianan Puffbird (*Notharchus macrorhynchos*)
Black Nunbird (*Monasa atra*)
Guianan Toucanet (*Selenidera piperivora*)
Green Araçari (*Pteroglossus viridis*)
Golden-collared Woodpecker (*Veniliornis cassini*)
Chestnut-rumped Woodcreeper (*Xiphorhynchus pardalotus*)
Black-throated Antshrike (*Frederickena viridis*)
Band-tailed Antshrike (*Sakesphorus melanothorax*)
Northern Slaty-Antshrike (*Thamnophilus punctatus*)
Guianan Streaked-Antwren (*Myrmotherula surinamensis*)
Rufous-bellied Antwren (*Myrmotherula guttata*)
Brown-bellied Antwren (*Epinecrophylla gutturalis*)
Todd's Antwren (*Herpsilochmus stictocephalus*)
Guianan Warbling-Antbird (*Hypocnemis cantator*)
Black-headed Antbird (*Percnostola rufifrons*)
Ferruginous-backed Antbird (*Myrmeciza ferruginea*)
Rufous-throated Antbird (*Gymnopithys rufigula*)
Boat-billed Tody-Tyrant (*Hemitriccus josephinae*)
Painted Tody-Flycatcher (*Todirostrum pictum*)
Capuchinbird (*Perissocephalus tricolor*)
Guianan Red-Cotinga (*Phoenicercus carnifex*)
Guianan Cock-of-the-Rock (*Rupicola rupicola*)
White-throated Manakin (*Corapipo gutturalis*)
White-fronted Manakin (*Lepidothrix serena*)
Tiny Tyrant-Manakin (*Tyranneutes virescens*)
Tepui Greenlet (*Hylophilus sclateri*)
Blue-backed Tanager (*Cyanicterus cyanicterus*)
Golden-sided Euphonia (*Euphonia cayennensis*)

downstream from camp. This species specializes on this type of bamboo, and is never found away from it; accordingly, it has a patchy distribution in Amazonia. Our records extend the known range of the species eastward by approximately 900 km; the nearest known locality for this species is along the Rio Siapa in Amazonas, Venezuela (Hilty 2003). This is the first record for any of the Guianas.

Populations of parrots, guans, and curassows in the Konashen COCA seemed healthy. Fourteen species of parrots were encountered, including *Ara macao* (Scarlet Macaw), a CITES I species, and Blue-cheeked Parrot (*Amazona dufresniana*), listed as Near Threatened (IUCN 2006). We did not encounter any other species considered to be of global conservation concern (IUCN), although *Harpia harpyja* (Harpy Eagle; IUCN Near Threatened) is well known to the Wai-Wai and undoubtedly occurs in the Konashen COCA. Guans and curassows were encountered frequently, particularly along the Kamoa River. They showed little fear of human observers, suggesting that hunting pressure on these species is relatively low.

DISCUSSION

The Konashen COCA has a rich avifauna. 319 species were encountered during our survey, and it is certain that at least 400 species of birds occur in the area. The avifauna was typical of a large region of undisturbed lowland forest in the Guayana Shield. The considerable habitat diversity within a relatively small area at each survey site was probably responsible for species lists that were higher than expected, given the short duration of the survey; however, we only found one species that was genuinely unexpected. Species endemic to the Guayana Shield were well represented in the Konashen COCA; again, virtually all of these were expected to occur at our survey sites, since they are widely distributed in lowland forests of the Guayana Shield.

The majority of birds observed in the Konashen COCA were found inside tall forest – either riverine (RF) or terra firme (TF) forest, or both (Appendix 3). Within these habitats, the bird community was dominated by suboscine passerines in the families Furnariidae (Ovenbirds), Thamnophilidae (Antbirds), and Tyrannidae (Tyrant Flycatchers), which collectively comprised almost one-third of all species recorded. Members of these families formed the core of mixed-species foraging flocks in the understory and canopy. Such flocks were commonly encountered in terra firme forest, where they were typically large and diverse; by contrast, mixed-species flocks were less frequent in riverine forest, especially where the understory was sparse and the forest relatively short in stature (as was the case in the immediate vicinity of the Kamoa River). Flocks in riverine forest tended to contain fewer species than those in terra firme.

Several species that are generally uncommon or local in the Guianas were more common in the Konashen COCA than we had expected. These species include *Touit purpuratus*

(Sapphire-rumped Parrotlet), *Xenops milleri* (Rufous-tailed Xenops), *Myrmotherula guttata* (Rufous-bellied Antwren), and *Tangara chilensis* (Paradise Tanager). Several other species, including *Heliornis fulica* (Sungrebe), *Florisuga mellivora* (White-necked Jacobin), *Hypocnemoides melanopogon* (Black-chinned Antbird), *Ramphotrigon ruficauda* (Rufous-tailed Flatbill), *Lathrotriccus euleri* (Euler's Flycatcher), and *Phoenicercus carnifex* (Guianan Red-Cotinga), appeared to be more common in the Konashen COCA during our survey than we have found them to be at other sites in Guyana and Suriname. Other rare or poorly known species that we encountered were *Leucopternis melanops* (Black-faced Hawk), *Nyctibius leucopterus* (White-winged Potoo), *Nyctibius brachteatus* (Rufous Potoo), *Dendrocincla merula* (White-chinned Woodcreeper, rare in Guyana and Suriname), *Sakesphorus melanothorax* (Band-tailed Antshrike), *Hemitriccus josephinae* (Boat-billed Tody-Tyrant), and *Neopipo cinnamomea* (Cinnamon Tyrant-Manakin). We expected to encounter the common and widespread *Piaya cayana* (Squirrel Cuckoo) and *Pachyramphus minor* (Pink-throated Becard) during this survey, but we found neither species. *Piculus flavigula* (Yellow-throated Woodpecker), usually a common member of canopy mixed-species flocks in the Guianas, was unaccountably scarce in the Konashen COCA during our survey.

Eighteen species were found primarily (or only) at and above 800 m during our survey. With the exception of *Hylophilus sclateri* (Tepui Greenlet), the highland avifauna that we could access from the Acarai site did not contain any species endemic to the Pantepui area. However, it is possible that populations of some highland species in the Acarai Mountains may be genetically distinct from Pantepui populations, from which they are isolated by extensive regions of lowland forest.

Although it contained few surprises, the avifauna of the Konashen COCA was representative of a spectacular, undisturbed tropical forest ecosystem, and its global conservation value cannot be overestimated. Birds are a comparatively well-known and easily surveyed group of organisms. The fact that we observed a new bird species for Guyana, despite previous fieldwork in the same area by other researchers, strongly suggests that the biodiversity of the Konashen COCA is greater than we could assess during this brief visit. We have no doubt that the species list for the Konashen COCA would continue to grow with further survey effort.

The parrot fauna of the Konashen COCA seemed not to be affected by the large-scale trapping of birds for the pet trade that plagues more accessible regions of the Guianas (Hanks 2005). Fourteen species were observed, many of which seemed to be in good numbers (with the caveat that parrot populations can fluctuate dramatically at a single location over the course of a year as the birds track the availability of their preferred foods). The Wai-Wai corroborated my impressions, stating that trappers do not visit the Konashen COCA, and that Wai-Wai only trap parrots occasionally for the purpose of keeping them as pets (rather than exporting them to coastal markets). The Wai-Wai do hunt some larger parrots, particularly the macaws (*Ara* spp.), but the effects of this hunting appear to be negligible.

Guans and curassows (Cracidae; hereafter "cracids") were seen frequently at both sites during the survey, suggesting that their populations are healthy; this was expected considering the remoteness of the Konashen COCA, the intact nature of the forests in the region, and the low human population density along the Guyana-Brazil border. In general, cracids have relatively low reproductive rates, are rather sluggish and easily shot, and are prized for food. They are notoriously vulnerable to habitat fragmentation, often being among the first species to disappear when humans move into an area. Cracids in the Konashen COCA are subjected to locally intense hunting pressure by the Wai-Wai, but their regional populations appear to be able to withstand the effects of such hunting. While cracid populations in the Konashen COCA are healthy and not immediately threatened, they are nevertheless of great value from the perspective of global conservation. However, current rates of harvesting by the Wai-Wai do not conflict with a global conservation strategy for these species.

CONSERVATION RECOMMENDATIONS

The remarkable avian diversity of the Konashen COCA is under little threat at the present time. However, its global significance as a large intact region of tropical forest should be recognized, and care taken to forestall declines in species that currently maintain healthier populations here than elsewhere in their ranges. The following guidelines should be adopted.

1. *The Wai-Wai should exclude illegal Brazilian gold miners from their territory.*
The greatest potential threat to the Konashen COCA is the ongoing construction of a highway across northern Brazil, which will likely exacerbate the current problems associated with illegal miners in the interior of the Guianas. If necessary, the Wai-Wai should enlist the assistance of the government of Guyana to keep illegal miners out of their territory.

2. *Continue to avoid trapping parrots for coastal markets and the international pet trade.*
The Guianas contribute a substantial number of parrots to the international pet trade, and trappers often travel great distances to harvest the most valuable species. This has led to dramatic declines in the populations of some species in accessible areas closer to the coastal plain than the Konashen COCA. The remoteness of the Konashen COCA has no doubt served to protect it from such exploitation.

The status of parrot populations can be difficult to assess, particularly during short surveys. Parrots rely on ephemeral resources and wander widely on a seasonal basis. The 14 species that we observed in the Konashen COCA are presumed to have healthy regional populations when all factors (isolation, low human population, extent of intact habitat) are taken into account. We draw these conclusions despite the fact that not all species seemed equally common during the survey. We therefore suggest that protection from a known threat (trappers) will be more effective than implementation of a monitoring program. Monitoring of parrots is likely to yield spurious and biased data. Parrots are not amenable to standard survey methods because their abundance in any given area can vary substantially over the course of a year. Such fluctuations are more likely due to local factors, rather than more significant regional population trends, especially in relatively pristine areas such as the Konashen COCA.

3. *Develop and implement a rotation system to distribute the effects of subsistence hunting over as large an area as possible.*
Cracids are arguably the most important birds in the diet of the Wai-Wai. They have low reproductive rates and tend to disappear when subjected to heavy hunting pressure. The cracid populations in the Konashen COCA are undoubtedly healthy, and it is likely that local population depletion (due to hunting) is a temporary phenomenon in most cases. Nevertheless, hunting of cracids should be done judiciously by distributing hunting activity over as large an area as possible, such that the majority of the Konashen COCA is not used for hunting at any given time. This simple system would ensure that local populations have time to recover following brief periods of intense hunting.

As is the case for parrots, cracids are difficult to survey using traditional methods. Sample sizes for each survey are likely to be low, since these birds tend to be relatively uncommon. As frugivores that wander widely on a seasonal basis, their local populations may vary in a manner that has no relevance to regional population trends. Data from cracid surveys would therefore be of limited value. Identifying and addressing the most significant current threat (hunting by inhabitants of the Konashen COCA) is the best conservation strategy at the present time.

REFERENCES

Braun, M.J., D.W. Finch, M.B. Robbins and B.K. Schmidt. 2000. A Field Checklist of the Birds of Guyana. Biological Diversity of the Guianas Program, Publ. 41, Smithsonian Institution, Washington, DC.

Braun, M.J., M.B. Robbins, C.M. Milensky, B.J. O'Shea, B.R. Barber, W. Hinds and W.S. Prince. 2003. New birds for Guyana from Mts. Roraima and Ayanganna. Bulletin of the British Ornithological Club 123: 24-33.

Finch, D.W, W. Hinds, J. Sanderson and O. Missa. 2002. Avifauna of the Eastern Edge of the Eastern Kanuku Mountains, Lower Kwitaro River, Guyana. pp. 43-46. *In:* Montambault, J.R. and O. Missa (eds.). A Biodiversity Assessment of the Eastern Kanuku Mountains, Lower Kwitaro River, Guyana. RAP Bulletin of Biological Assessment 26. Conservation International, Washington, DC.

Hanks, C.K. 2005. Spatial patterns in Guyana's wild bird trade. MA Thesis, University of Texas, Austin.

Hilty, S. L. 2003. Birds of Venezuela. 2nd Ed. Princeton, NJ: Princeton University Press.

Naka, L.N., M. Cohn-Haft, F. Mallet-Rodrigues, M.P.D. Santos and M.F.M. Torres. 2006. The avifauna of the Brazilian state of Roraima: bird distribution and biogeography in the Rio Branco basin. Revista Brasileira de Ornitologia 14(3): 197-238.

O'Shea, B.J., C.M. Milensky, S. Claramunt, B.K. Schmidt, C.A. Gebhard, C.G. Schmitt and K.T. Erskine. 2007. New records for Guyana, with description of the voice of Roraiman Nightjar (*Caprimulgus whitelyi*). Bulletin of the British Ornithological Club 127: 118-128.

Parker, T.A., III, R.B. Foster, L.H. Emmons, P. Freed, A.B. Forsyth, B. Hoffman and B.D. Gill (eds.). 1993. A biological assessment of the Kanuku Mountain region of southwestern Guyana. RAP Working Papers 5. Conservation International, Washington, DC.

Ridgely, R.S., D. Agro and L. Joseph. 2005. Birds of Iwokrama Forest. Proceedings of the Academy of Natural Sciences 154:109-121.

Robbins, M.B., M.J. Braun, and D.W. Finch. 2004. Avifauna of the Guyana southern Rupununi, with comparisons to other savannas of northern South America. Ornitologia Neotropical 15:173-200.

Robbins, M.B., M.J. Braun, C.M. Milensky, B.K. Schmidt, W. Prince, N.H. Rice, D.W. Finch, and B.J. O'Shea. 2007. Avifauna of the upper Essequibo River and Acary Mountains, southern Guyana. Ornitologia Neotropical 18:339-368.

Chapter 8

Non-Volant Mammals of the Konashen COCA, Southern Guyana

*James G. Sanderson, Eustace Alexander,
Vitus Antone and Charakura Yukuma*

SUMMARY

We present the results of a large non-volant mammal survey conducted during a Rapid Assessment Program (RAP) expedition at two sites in the Konashen Community Owned Conservation Area (COCA) of southern Guyana from October 4 - 27, 2006. The purpose of the survey was to assess and document the biological diversity of large mammals and use the results to guide the development of a conservation management plan for sustainable resource utilization by inhabitants of the area. To survey for the presence of large mammals we used three methodologies: (1) tracks, scats, sounds, and visual observations (including hand-held photographs), (2) interviews with local people, and (3) camera phototraps. We suspect the presence of 42 large mammal species and confirmed 21 in the region. According to the 2008 IUCN Red List the Brown-bearded saki monkey (*Chiropotes satanas*) and the Giant otter (*Pteronura brasiliensis*) are listed as Endangered, and the Giant armadillo (*Priodontes maximus*), Bush dog (*Speothos venaticus*) and Brazilian tapir (*Tapirus terrestris*) are considered Vulnerable. Both study sites are utilized as hunting areas for two weeks per year by the local people, but were otherwise pristine, undisturbed tropical rain forest. Our evidence suggests that the sites we sampled contain the full complement of the large mammal species characteristic of the Guayana Shield. Because this region has a very low human population density (0.032 humans/km²) the forests of the Konashen COCA is likely to contain an intact faunal assemblage of large mammals.

INTRODUCTION

To implement effective conservation strategies, information on specific local biological diversity is essential. Often such information is unknown, incomplete, or unavailable to policymakers. The large mammalian fauna of the Guayana Shield Region is well known and widely distributed and, because the human population density is low, few species are severely threatened. Though much of the region is unoccupied, some areas support small numbers of Amerindian communities. Residents of these communities are mostly subsistence hunters that clear out small areas for cultivation. The region of southern Guyana such as the Konashen Indigenous District that borders Brazil to the south is an example.

The Konashen Indigenous District (KID) is approximately 625,000 hectares and is legally owned by fewer than 200 Amerindian individuals, mainly of Wai-Wai ancestry, living in the single community of Masakenari (labeled as Konashen on typical maps of Guyana) along the Essequibo River, and just north of the confluence of the Essequibo and Kamoa rivers. Small areas near the village have been cleared for cultivation. During the wet season the nearby landing strip at Gunns (located south of the confluence of the Essequibo and Kassikaityu rivers) in the savanna is flooded, preventing access by air to the village. Masakenari consists of single family houses, a school house, and a community center built of wood on a hill overlooking the Essequibo River. During the wet season the flooded river can rise more than 6 m. The previous village – Akuthopono – approximately 6 km distant, was flooded in 2000, and forced the establishment of Masakenari.

Since little is known of the local biodiversity in many of these remote regions, and because conservation programs include increasing local awareness and interest in biodiversity, a RAP survey of birds, amphibians and reptiles, insects, fishes, and large mammals was undertaken. The objective of the RAP survey in the forests of the Konashen COCA was to provide quick, efficient, reliable, and cost-effective biodiversity data at two sites to support local, national and regional conservation strategies.

MATERIALS AND METHODS

Study Area

We conducted our surveys in the dry season at two sites in the Konashen COCA of southern Guyana (N 1°25', W 58°57') from 4-26 October, 2006. The Konashen COCA is the most southerly area of Guyana and borders the Brazilian state of Para along the Acarai Mountains, an arc of mountains running east-west that separates Guyana from Brazil. The Wassarai Mountains and the Kamoa Mountains are located north and west of our study areas. Henceforth, we refer to our study sites as Acarai (N 1°22'59.4", W 58°56'49.2"), and Kamoa (N 1° 31'52.0", W 58°49',40.2"). The elevation range surveyed was 230 - 1300 m, and 250 – 512 m, for Acarai and Kamoa, respectively. The Acarai study site was located on a tributary of the Sipu River, and the Kamoa study site was located along the main Kamoa River. Both the Sipu and the Kamoa rivers form part of the Essequibo River headwaters.

We include here the results of two other camera trapping efforts also performed in the Konashen Indigenous District: the first during February 2006 at Wanakoko (N 1°41'46.3", W 58°38'52.4") located 15 km north of Masakenari along the Essequibo River, and the second along the old trail to Suriname (located at N 1°23'52.8", W 58°22'14"). In the greater Konashen Indigenous District the forest below approximately 250 m was seasonally inundated. At Acarai our camera traps were located in the foothills of the Acarai Mountains at an elevation of approximately 350 m, and at Kamoa our camera traps were in seasonally flooded forest.

METHODS

To survey for the presence of large non-volant mammals we used three methodologies: (1) tracks, scats, sounds, and visual observations supplemented with hand-held photographs when possible, (2) interviews with local people, and (3) camera phototraps at each of the two study sites. Because these methods provide different confidence levels all results are presented separately. To determine the presence of large non-volant mammalian species, we recorded direct observation of species, track and sound identification, nests, dung and other indirect information made each day during daily excursions from base camp. Because our records were also collected opportunistically by our colleagues, and some observations may have been repeated, we used this information only to document species presence.

Interviews of local people were conducted using *Neotropical Rainforest Mammals* (Emmons 1999) as a guide. During interviews, individuals were asked to page through the book and identify the photos of mammals they had observed in the recent past in their forest. We avoided making comments that might influence their decisions, and no time pressure was used to coerce responses.

For shy mammals experiencing hunting pressure, camera trapping methods might be more effective than walking transects, especially when observers have different and varied levels of expertise. Camera trap photographs also provide direct evidence for species presence because the pictures are available for anyone to review. The passive method included the use of eight camera phototraps (Trapa-camera, Saõ Paulo, Brazil) operated at each study site. Trapa-camera photo traps are triggered by heat-in-motion and operate with a standard 35 mm camera set on autofocus, loaded with ASA 200 print film, and powered by 5 AA batteries. Cameras were set to operate continuously and to wait approximately 10 seconds between photographs. Cameras were placed at sites suspected of being frequented by various mammalian species. Den sites, trails, feeding or drinking stations, and fallen trees across streams were chosen for camera placement. Cameras were typically located at least 500 m from base camp. Animals were attracted to the camera traps with Hawbaker's Wild Cat Lure #2 (Adirondack Outdoor Company, Elizabethtown, NY USA).

RESULTS

We observed, identified by tracks, scats, or sound 17 and 19 species of large mammals at Acarai and Kamoa, respectively (Appendix 4), yielding a combined total of 22 species of large mammals. Interviews with local people revealed the possible presence of a total of 42 species of large mammals in Konashen COCA including several species such as the Crab-eating fox (*Cerdocyon thous*) and White-tailed deer (*Odocoileus virginianus*), found mostly in savanna (Appendix 4). A total of 3 and 2 large mammals were photographed by our camera traps at Acarai and Kamoa, respectively, during 18 and 12 camera-trap days, respectively. Our camera traps at Wanakoko photographed 7 mammal species during 32 days of operation. Our local guides also provided direct evidence of species' presences.

DISCUSSION

Our results suggest that the full biologically rich assortment of large mammals, which characterize the Guayana Shield, remains intact within the Konashen COCA. Because the human population is very low in the Konashen Indigenous

District, hunting pressure is unlikely to have any significant impact on the large mammals.

According to the 2008 IUCN Red List the majority of large mammals we documented to occur were not threatened. However, the Brown-bearded saki monkey (*Chiropotes satanas*) and the Giant otter (*Pteronura brasiliensis*) are listed as Endangered, the Giant armadillo (*Priodontes maximus*), Bush dog (*Speothos venaticus*), and Brazilian tapir (*Tapirus terrestris*) are considered Vulnerable, and the Giant anteater (*Myrmecophaga tridactyla*), the Oncilla or Tigrina (*Leopardus tigrinus*), Puma (*Puma concolor*), and Jaguar (*P. onca*) are Near Threatened. However, these large mammal species, that occur throughout the Guayana Shield, are relatively secure in this area given the low human population density throughout the region.

Because so few camera trap pictures of large animals were obtained, a discussion of photographic rates is not useful. However, the fact that so few photographs were obtained suggests that wildlife occurs in low densities or is extremely shy at the sites we surveyed. We suspect that seasonally flooded forests are less than optimal habitat for territorial, large mammals.

Local pressure on forest resources for fuel wood, building of homes, hunting, and clearing forest patches for cultivation is minimal. With fewer than 200 occupants living within 625,000 ha and readily available fish and birds as alternate sources of protein, the large mammalian fauna is secure. Unlike other intact forests in some regions of the world, our results suggest that the so-called *empty forest syndrome* (Redford 1992) does not occur and, moreover, is not in danger of occurring. Our results show that an intact faunal assemblage of large mammals is secure in the Konashen COCA of southern Guyana.

CONSERVATION STATUS AND ACTION RECOMMENDATIONS

Because the human density is low and pressure on natural resources is carefully managed by community leaders, the intact faunal assemblage of large mammals typical of the Guayana Shield remains secure in the Konashen Indigenous District.

The greatest threat to biodiversity is likely to come from external sources far beyond the Wai-Wai community. The task of Wai-Wai community leadership is to manage their resources in a sustainable manner and prevent outsiders from jeopardizing the ecological integrity of the area.

Though the intact faunal assemblages of large mammals appear to be secure, the community needs to implement and manage a long-term monitoring program to detect any changes in the occurrence of species, especially those that are listed by IUCN as Endangered.

REFERENCES

Boufford, D.E., P.P. van Dijk, and L. Zhi. 2004. Mountains of Southwest China. Pages 159 – 164 *in* Mittermeier, R.A., P. Robles Gil, M. Hoffmann, J. Pilgrim, T. Brooks, C.G. Mittermeier, J. Lamoreux, G. A.B. da Fonseca, eds. Hotspots Revisited. Cemex.

Emmons, L.H. 1999. Neotropical Rainforest Mammals, University of Chicago Press, Chicago, USA.

Redford, K.H. 1992. The Empty Forest. *BioScience* 42(6): 412-422.

Appendix 1

Preliminary list of ant genera collected in the Konashen COCA, October 2006

Ted R. Schultz and Jeffrey Sosa-Calvo

List of ant genera collected from the samples with estimates of species diversity in parentheses (no numbers indicates still processing or uncertain)

Amblyoponinae	**Dolichoderinae**
Prionopelta (2)	*Azteca* (3)
	Dolichoderus (3)
Ecitoninae	
Eciton (1)	**Pseudomyrmicinae**
Labidus (1)	*Pseudomyrmex* (1)
Ectatomminae	**Myrmicinae**
Ectatomma (2)	*Acromyrmex*
Gnamptogenys (14)	*Apterostigma*
	Atta
Formicinae	*Cephalotes*
Acropyga (6)	*Crematogaster* (10)
Camponotus (5)	*Cyphomyrmex*
Gigantiops (1)	*Daceton* (1)
Paratrechina (7)	*Mycetarotes*
	Mycocepurus
	Myrmicocrypta
Paraponerinae	*Pheidole* (ca. 50)
Paraponera (1)	*Procryptocerus*
	Rogeria (3)
Ponerinae	*Solenopsis*
Anochetus (3)	*Strumigenys* (11)
Hypoponera (11)	*Trachymyrmex*
Leptogenys (2)	*Wasmannia* (2)
Odontomachus (3)	
Pachycondyla (5)	

Appendix 2

Abundance matrix of fish species collected during the 2006 RAP survey in the Acarai Mountains, Sipu, Kamoa and Essequibo rivers, Konashen Indigenous District of Southern Guyana

Carlos A. Lasso, Jamie Hernández-Acevedo, Eustace Alexander, Josefa C. Señaris, Lina Mesa, Hector Samudio, Antoni Shushu, Elisha Mauruwanaru and Romel Shoni

* Species observed but not collected.

	SR-01	SR-02	SR-03	SR-04	SR-05	AM-06	AM-07A	AM-07B	SR-08	KR-09	KR-10	KR-11	KR-12	WL-13	PS-14-a	PS-14-b	PF-15	MAKR-16	Total
CHARACIFORMES (61)																			
Acestrorhynchidae (2)																			
Acestrorhynchus falcatus	1			1	4					1									7
Acestrorhynchus microlepis	3			1						1				1					6
Anostomidae (5)																			
Anostomus anostomus	1		1					2								1		1	6
Leporinus arcus	1							1											2
Leporinus nigrotaeniatus																		1	1
Leporinus gr. *maculatus*								4								1		1	6
Schizodon fasciatum														1					1
Characidae (31)																			
Aphyocharax erythrurus																2			2
Brachychalcinus orbicularis	1									1				1		3			6
Bryconops affinis																1			1
Bryconops caudomaculatus	3							3		2		6		1		5		1	21
Chalceus macrolepidotus	1	1																	2
Creagrutus melanzonus								1											1
Charax gibbosus	2	1																	3
Cynopotamus essequibensis														1					1
Hemigrammus guayanensis	3				4						23	30	8	25	36				129
Hemigrammus unilineatus								4	78										82
Hemigrammus sp.					1				46		47	40		57	109	16			316
Hyphessobrycon bentosi			80	71	2						134	82	23	169	73	46			680
Hyphessobrycon minor			4	9								2	6	49					70
Jupiaba abramoides	1				8			4	1	1									15
Jupiaba pinnata																		1	1
Jupiaba polylepis	1		1	11				7				1		1				7	29
Jupiaba potaroensis							1		8										9
Moenkhausia collettii			120	60				4	53		24	26	8			11			306
Moenkhausia oligolepis				2	7														9
Moenkhausia grandisquamis								2		1									3
Moenkhausia gr. *lepidura*														3					3
Moenkhausia lepidura lepidura	2	6						2		1	2	5		1					19
Moenkhausia lepidura lata				20															20
Poptella longipinnis														1					1
Myleus rhomboidalis										1									1
Serrasalmus rhombeus	1									1				1					3

Abundance matrix of fish species collected during the 2006 RAP survey in the Acarai Mountains, Sipu, Kamoa and Essequibo rivers, Konashen Indigenous District of Southern Guyana

	SR-01	SR-02	SR-03	SR-04	SR-05	AM-06	AM-07A	AM-07B	SR-08	KR-09	KR-10	KR-11	KR-12	WL-13	PS-14-a	PS-14-b	PF-15	MAKR-16	Total
Tetragonopterus chalceus																1			1
Phenacogaster microstictus			6	19					16		18	5							64
Phenacogaster megalostictus																8			8
Triportheus brachipomus										1				1					2
Triportheus sp.																		1	1
Crenuchidae (5)																			
Amnocryptocharax lateralis			6															15	21
Amnocryptocharax vintoni								1											1
Characidium steindachneri			2					2			2	2	2					2	12
Melanocharacidium blennioides								1			1							2	4
Characidium sp.																		2	2
Curimatidae (5)																			
Curimatella dorsalis	1												3	1					5
Curimata cyprinoides														1					1
Cyphocharax sp.				1							6	4				1			12
Steindachnerina sp.	1																		1
Curimatopsis crypticus															2				2
Cynodontidae (3)																			
Cynodon gibbus*																			0
Hydrolycus armatus*																			0
Hydrolycus tatauaia*																			0
Erythrinidae (4)																			
Hoplias macrophthalmus				1	2			1						1		1		1	7
Hoplias cf malabaricus					2				1	1	2	1			1				8
Hoplerythrinus unitaeniatus																	2		2
Erythrinus erythrinus					1				1								1		3
Hemiodontidae (2)																			
Hemiodus vorderwinkleri																		2	2
Hemiodus semitaeniatus	2									1				1				1	5
Lebiasinidae (2)																			
Nannostomus marginatus			18		1						2	6		1	303	1			332
Pyrrhulina filamentosa			3	4	31				7		1	1	1	14	9	2			73
Parodontidae (1)																			
Parodon guyanensis								2								8			10
Prochilodontidae (1)																			
Prochilodus rubrotaeniatus*																			0
SILURIFORMES (32)																			

	SR-01	SR-02	SR-03	SR-04	SR-05	AM-06	AM-07A	AM-07B	SR-08	KR-09	KR-10	KR-11	KR-12	WL-13	PS-14-a	PS-14-b	PF-15	MAKR-16	Total
Auchenipteridae (4)																			
Ageneiosus marmoratus										1				1					2
Auchenipterus demerarae										1									1
Tatia intermedia	1	1	3									1							6
Trachyliopterus galeatus															1		1		2
Callichthyidae (4)																			
Corydoras bondi				2				4											6
Corydoras sp.																1			1
Callichthys callichthys					2	1													3
Megalechis thoracata																	8		8
Cetopsidae (2)																			
Helogenes marmoratus								2	1										3
Pseudocetopsis macilenta		2																	2
Doradidae (1)																			
Doras cf. *micropoeus*														2					2
Heptapteridae (5)																			
Pimelodella cf. *gracilis*						2				1		1				1			5
Rhamdia cf. *foina*												1	1						2
Pimelodella cristata		1								1									2
Imparfinis cf. *pijpersi*																		1	1
Pimelodella sp.										1									1
Loricariidae (13)																			
Ancistrus cf. *lithurgicus*																1			1
Ancistrus sp. 1				1				10											11
Ancistrus sp. 2		3	8					1											12
Hypostomus cf. *hemiurus*																2		4	6
Chasmocranus longinor								6										3	9
Harttia sp.								4										7	11
Hemiancistrus sp.																		1	1
Hypostomus taphorni	1									2						1			4
Hypostomus sp.								14								1			15
Lithoxus cf. *surinamensis*								1											1
Pseudoancistrus barbattus																1			1
Pseudoancistrus nigrescens																		1	1
Rineloricaria platyura			1					2											3
Pimelodidae (2)																			
Pimelodus ornatus										1									1
*Pseudoplatystoma fasciatum**																			0
Trichomycteridae (1)																			

Abundance matrix of fish species collected during the 2006 RAP survey in the
Acarai Mountains, Sipu, Kamoa and Essequibo rivers, Konashen Indigenous
District of Southern Guyana

	SR-01	SR-02	SR-03	SR-04	SR-05	AM-06	AM-07A	AM-07B	SR-08	KR-09	KR-10	KR-11	KR-12	WL-13	PS-14-a	PS-14 b	PF-15	MAKR-16	Total
Ochmacanthus flabelliferus												1				1			2
GYMNOTIFORMES (9)																			
Gymnotidae (3)																			
Gymnotus carapo					1										7		1		9
Gymnotus coropinae				2							1		2						5
*Electrophorus electricus**																			0
Hypopomidae (2)																			
Brachyhypopomus brevirostris					8								2		2		1		13
Hypopygus lepturus											21	1			14				36
Rhamphichthydae (1)																			
Gymnorhamphichthys petitti											14								14
Sternopygidae (3)																			
Eigenmannia humboldtii														1					1
Eigenmannia virescens															4				4
Sternopygus macrurus								2											2
CYPRINODONTIFORMES (1)																			
Cyprinodontidae (1)																			
Rivulus sp.							10												10
PERCIFORMES (9)																			
Cichlidae (8)																			
Aequidens tetramerus	1			1	10					2				1		1	1		17
Bujurquina sp.								9											9
Apistogramma steindachneri			1	40	7				1		5	2	2	4					62
Crenicichla alta	1			3				1				1	3						9
Crenicichla cf. *lenticulata*														1					1
Crenicichla lugubris														2					2
Guianacara sp.		2		23														1	26
Geophagus surinamensis														1					1
Sciaenidae (1)																			
*Petilipinnis grunniens**																			0
SYNBRANCHIFORMES (1)																			
Synbranchidae (1)																			0
Synbranchus marmoratus	1										1								2
Total	31	17	254	272	91	3	11	97	213	23	304	217	61	345	561	118	15	57	2690

Appendix 3

Preliminary Bird Species Checklist of the Konashen COCA, Southern Guyana

Brian J. O'Shea

Birds observed in the Konashen COCA between October 6th and 28th, 2006. Taxonomy, nomenclature, and order of the list follow the most current information provided by the American Ornithologists' Union South American Checklist Committee (Remsen et al. 2008). Abundance codes are given in the Methods section (Chapter 7). Abundances refer to the habitat in which a given species is most common.

Species that were only seen while traveling between the Acarai and Kamoa sites are listed in the "Transit" column. Habitat codes are as follow:

TF: terra firme; tall forest on well-drained soil, never inundated, up to 800 m
RF: riverine forest; seasonally inundated, sparse understory, many lianas
RI: immediate vicinity of rivers; on or over water, at edge or close to it (Kamoa only)
MT: montane forest; above 800 m (Acarai only)
SF: savanna forest (seasonally inundated, < 300 m, "palm swamp") (Kamoa only)
SG: secondary growth along creeks, dominated by *Guadua* bamboo (Acarai only)
OV: overhead; no specific habitat

REFERENCES

Remsen, J.V., Jr., C.D. Cadena, A. Jaramillo, M. Nores, J.F. Pacheco, M.B. Robbins, T.S. Schulenberg, F.G. Stiles, D.F. Stotz and K.J. Zimmer. 2008. A classification of the bird species of South America. American Ornithologists' Union. http://www.museum.lsu.edu/~Remsen/SACCBaseline.html

Scientific name	English common name	Acarai	Kamoa	Transit	Habitat	Abundance	Elevation
Tinamidae	**Tinamous**						
Tinamus major	Great Tinamou	X	X		TF, RF	F	
Crypturellus cinereus	Cinereous Tinamou	X	X		RF	U	
Crypturellus soui	Little Tinamou	X			MT	U	
Crypturellus undulatus	Undulated Tinamou		X		RF, SF	U	
Crypturellus variegatus	Variegated Tinamou	X	X		TF, RF	F	
Crypturellus brevirostris	Rusty Tinamou	X			TF	U	
Anatidae	**Ducks, Geese**						
Cairina moschata	Muscovy Duck		X		RI	U	
Cracidae	**Curassows, Guans**						
Ortalis motmot	Variable Chachalaca	X	X		RF, TF, SG, SF	F	
Penelope jacquacu	Spix's Guan	X	X		RF, TF, MT	C	
Pipile cumanensis	Blue-throated Piping-Guan	X	X		RF	F	
Crax alector	Black Curassow	X	X		RF, TF, MT	F	
Odontophoridae	**Quails**						

Scientific name	English common name	Acarai	Kamoa	Transit	Habitat	Abundance	Elevation
Odontophorus gujanensis	Marbled Wood-Quail	X	X		TF, RF	F	
Anhingidae	**Anhingas**						
Anhinga anhinga	Anhinga		X		RI	C	
Ardeidae	**Herons**						
Zebrilus undulatus	Zigzag Heron	X	X		TF, RF	U	
Tigrisoma lineatum	Rufescent Tiger-Heron		X		RI	F	
Ardea cocoi	Cocoi Heron		X		RI	U	
Butorides striata	Striated Heron		X		RI	U	
Agamia agami	Agami Heron		X		RI	U	
Ardea alba	Great Egret			X	RI		
Egretta caerulea	Little Blue Heron			X	RI		
Threskiornithidae	**Ibises**						
Mesembrinibis cayennensis	Green Ibis	X	X		RI, RF	F	
Cathartidae	**Vultures**						
Sarcoramphus papa	King Vulture	X	X		OV	U	
Cathartes melambrotus	Greater Yellow-headed Vulture	X	X		OV	F	
Pandionidae	**Ospreys**						
Pandion haliaetus	Osprey		X		RI	U	
Accipitridae	**Hawks, Eagles**						
Chondohierax uncinatus	Hook-billed Kite		X		OV	U	
Leucopternis albicollis	White Hawk		X		RI	U	
Leucopternis melanops	Black-faced Hawk	X			RF	U	
Buteogallus urubitinga	Great Black Hawk	X	X		OV, RI	U	
Buteo brachyurus	Short-tailed Hawk	X	X		OV	U	
Spizaetus tyrannus	Black Hawk-Eagle		X		RF	U	
Spizaetus ornatus	Ornate Hawk-Eagle	X	X		OV	U	
Falconidae	**Falcons, Caracaras**						
Daptrius ater	Black Caracara			X	RI		
Ibycter americanus	Red-throated Caracara	X	X		TF, RF	F	
Micrastur ruficollis	Barred Forest-falcon		X		RF	U	
Micrastur gilvicollis	Lined Forest-falcon	X	X		TF, RF	F	
Micrastur mirandollei	Slaty-backed Forest-falcon	X	X		RF, TF	F	
Falco rufigularis	Bat Falcon		X		OV	U	
Heliornithidae	**Sungrebes**						
Heliornis fulica	Sungrebe		X		RI	F	
Eurypygidae	**Sunbitterns**						
Eurypyga helias	Sunbittern	X	X		RI	U	
Psophiidae	**Trumpeters**						
Psophia crepitans	Gray-winged Trumpeter	X	X		TF, RF	U	
Scolopacidae	**Sandpipers**						
Tringa solitaria	Solitary Sandpiper			X			
Actitis macularia	Spotted Sandpiper		X		RI	U	
Columbidae	**Pigeons, Doves**						
Patagioenas subvinacea	Ruddy Pigeon	X	X		RF, TF	F	
Patagioenas plumbea	Plumbeous Pigeon	X	X		RF, TF	F	
Leptotila rufaxilla	Gray-fronted Dove	X	X		RF	U	
Geotrygon montana	Ruddy Quail-Dove	X			RF	U	
Psittacidae	**Parrots**						
Ara ararauna	Blue-and-yellow Macaw	X	X		RF, OV	F	
Ara macao	Scarlet Macaw	X	X		RF, OV	U	

Scientific name	English common name	Acarai	Kamoa	Transit	Habitat	Abundance	Elevation
Ara chloropterus	Red-and-green Macaw	X	X		RF, TF, MF, OV	F	
Aratinga leucophthalmus	White-eyed Parakeet	X	X		OV	U	
Pyrrhura picta	Painted Parakeet	X	X		RF	U	
Brotogeris chrysoptera	Golden-winged Parakeet	X	X		RF	U	
Touit purpuratus	Sapphire-rumped Parrotlet	X	X		RF	F	
Pionites melanocephalus	Black-headed Parrot		X		RF, SF	U	
Pionopsitta caica	Caica Parrot	X	X		RF, TF	C	
Pionus menstruus	Blue-headed Parrot	X	X		RF, TF, OV	F	
Pionus fuscus	Dusky Parrot	X	X		RF, TF, OV	F	
Amazona dufresniana	Blue-cheeked Parrot		X		RF	U	
Amazona amazonica	Orange-winged Parrot	X	X		RF, TF	F	
Deroptyus accipitrinus	Red-fan Parrot	X	X		RF, TF	U	
Cuculidae	**Cuckoos**						
Piaya melanogaster	Black-bellied Cuckoo	X	X		RF, TF	U	
Dromococcyx pavoninus	Pavonine Cuckoo	X	X		TF	U	
Neomorphus rufipennis	Rufous-winged Ground-Cuckoo	X			TF	U	
Strigidae	**Typical Owls**						
Megascops watsonii	Tawny-bellied Screech-Owl	X			RF	U	
Megascops guatemalae	Vermiculated Screech-Owl	X			MT	U	
Glaucidium hardyi	Amazonian Pygmy-Owl	X	X		RF, TF	U	
Lophostrix cristata	Crested Owl	X	X		RF, TF	U	
Pulsatrix perspicillata	Spectacled Owl		X		RF	U	
Nyctibiidae	**Potoos**						
Nyctibius griseus	Common Potoo	X	X		RF, TF	F	
Nyctibius leucopterus	White-winged Potoo		X		RF	U	
Nyctibius bracteatus	Rufous Potoo		X		RF	U	
Caprimulgidae	**Nighthawks, Nightjars**						
Lurocalis semitorquatus	Semicollared Nighthawk		X		RF	U	
Caprimulgus nigrescens	Blackish Nightjar	X	X		RI, SF	U	
Apodidae	**Swifts**						
Streptoprocne zonaris	White-collared Swift	X	X		OV	U	
Cypseloides cryptus	White-chinned Swift	X			OV, MT	U	
Chaetura chapmani	Chapman's Swift	X	X		OV	C	
Chaetura spinicaudus	Band-rumped Swift	X	X		OV	C	
Aeronautes montivagus	White-tipped Swift	X			OV, MT	C	
Panyptila cayennensis	Lesser Swallow-tailed Swift		X		OV, RI	U	
Tachornis squamata	Fork-tailed Palm-Swift			X	OV, RI		
Trochilidae	**Hummingbirds**						
Phaethornis superciliosus	Eastern Long-tailed Hermit	X	X		RF, TF, SG	F	
Phaethornis bourcieri	Straight-billed Hermit	X	X		RF, TF, SF, SG, MT	C	
Phaethornis ruber	Reddish Hermit	X	X		RF, TF, SF, SG	C	
Campylopterus largipennis	Gray-breasted Sabrewing	X	X		RF, TF	U	
Florisuga mellivora	White-necked Jacobin	X	X		RI, RF	C	
Colibri delphinae	Brown Violetear	X			MT	C	
Topaza pella	Crimson Topaz	X	X		RI, RF	F	
Lophornis ornatus	Tufted Coquette			X	RI		
Thalurania furcata	Fork-tailed Woodnymph	X	X		TF, RF, SG, MT	F	
Hylocharis sapphirina	Rufous-throated Sapphire		X		RF, SF	F	

Scientific name	English common name	Acarai	Kamoa	Transit	Habitat	Abundance	Elevation
Amazilia versicolor	Versicolored Emerald		X		RI	U	
Amazilia viridigaster	Green-bellied Emerald	X			MT	U	
Heliothryx auritus	Black-eared Fairy	X	X		RF, TF, SG	U	
Trogonidae	**Trogons**						
Trogon melanurus	Black-tailed Trogon	X	X		TF, RF	F	
Trogon viridis	White-tailed Trogon	X	X		RF, TF, SG	F	
Trogon collaris	Collared Trogon	X			MT	F	min elev 540 m
Trogon rufus	Black-throated Trogon	X	X		RF	U	
Trogon violaceus	Violaceous Trogon	X	X		RF, SG	F	
Alcedinidae	**Kingfishers**						
Megaceryle torquata	Ringed Kingfisher		X		RI	C	
Chloroceryle amazona	Amazon Kingfisher		X		RI	C	
Chloroceryle americana	Green Kingfisher		X		RI	C	
Chloroceryle inda	Green-and-rufous Kingfisher	X	X		RF, RI, SG	F	
Chloroceryle aenea	Pygmy Kingfisher	X	X		RF, RI, SG	F	
Momotidae	**Motmots**						
Momotus momota	Blue-crowned Motmot	X	X		RF, SG, TF	F	
Galbulidae	**Jacamars**						
Galbula albirostris	Yellow-billed Jacamar	X			TF	F	
Galbula leucogastra	Bronzy Jacamar		X		SF, RI	F	
Galbula dea	Paradise Jacamar	X	X		RF, TF, RI	C	
Jacamerops aureus	Great Jacamar	X	X		TF, RF	U	
Bucconidae	**Puffbirds**						
Notharchus macrorhynchos	Guianan Puffbird		X		RF	U	
Bucco tamatia	Spotted Puffbird	X	X		SF, RI	F	
Bucco capensis	Collared Puffbird	X			TF	U	
Malacoptila fusca	White-chested Puffbird	X			TF	U	
Nonnula rubecula	Rusty-breasted Nunlet		X		RF	U	
Monasa atra	Black Nunbird	X	X		RF, TF	F	
Chelidoptera tenebrosa	Swallow-winged Puffbird			X	RI		
Capitonidae	**New World Barbets**						
Capito niger	Black-spotted Barbet	X	X		RF, TF	U	
Ramphastidae	**Toucans**						
Aulacorhynchus derbianus	Chestnut-tipped Toucanet	X			MT	F	
Selenidera culik	Guianan Toucanet	X	X		TF, RF	U	
Pteroglossus aracari	Black-necked Araçari	X	X		RF	U	
Pteroglossus viridis	Green Araçari	X	X		RF, TF	U	
Ramphastos vitellinus	Channel-billed Toucan	X	X		TF, RF, MT	C	
Ramphastos tucanus	White-throated Toucan	X	X		TF, RF, MT	C	
Picidae	**Woodpeckers**						
Picumnus exilis	Golden-spangled Piculet			X	RI		
Piculus rubiginosus	Golden-olive Woodpecker	X			MT	F	
Piculus flavigula	Yellow-throated Woodpecker		X		RF	U	
Celeus elegans	Chestnut Woodpecker	X	X		RF, TF	U	
Celeus undatus	Waved Woodpecker	X	X		RF, TF	F	
Celeus flavus	Cream-colored Woodpecker	X	X		RF	U	
Celeus torquatus	Ringed Woodpecker		X		RF	U	
Dryocopus lineatus	Lineated Woodpecker	X			RF	U	

Scientific name	English common name	Acarai	Kamoa	Transit	Habitat	Abundance	Elevation
Veniliornis cassini	Golden-collared Woodpecker	X	X		TF, RF, SG, MT	C	
Campephilus melanoleucos	Crimson-crested Woodpecker	X			RF	F	
Campephilus rubricollis	Red-necked Woodpecker	X	X		TF, RF, MT	F	
Furnariidae	**Ovenbirds**						
Synallaxis rutilans	Ruddy Spinetail	X			RF	U	
Philydor pyrrhodes	Cinnamon-rumped Foliage-gleaner		X		RF	U	
Philydor ruficaudatum	Rufous-tailed Foliage-gleaner	X			TF	U	
Philydor erythrocercum	Rufous-rumped Foliage-gleaner	X			TF	F	
Automolus infuscatus	Olive-backed Foliage-gleaner	X	X		TF, RF	F	
Automolus ochrolaemus	Buff-throated Foliage-gleaner	X	X		RF, TF	F	
Automolus rufipileatus	Chestnut-crowned Foliage-gleaner	X			SG, RF	F	
Automolus rubiginosus	Ruddy Foliage-gleaner	X			SG	U	
Xenops minutus	Plain Xenops	X	X		TF, RF	F	
Xenops milleri	Rufous-tailed Xenops	X	X		TF, RF	U	
Sclerurus sp.	Leaftosser sp.	X	X		TF, RF	U	
Dendrocincla fuliginosa	Plain-brown Woodcreeper	X	X		TF, RF	C	
Dendrocincla merula	White-chinned Woodcreeper		X		RF	U	
Glyphorhynchus spirurus	Wedge-billed Woodcreeper	X	X		TF, RF, MT	C	
Sittasomus griseicapillus	Olivaceous Woodcreeper		X		SF	U	
Dendrexetastes rufigula	Cinnamon-throated Woodcreeper	X			RF	U	
Hylexetastes perrotii	Red-billed Woodcreeper	X			TF	U	
Xiphocolaptes promeropirhynchus	Strong-billed Woodcreeper		X		RF	U	
Dendrocolaptes certhia	Amazonian Barred-Woodcreeper		X		TF, RF	U	
Dendrocolaptes picumnus	Black-banded Woodcreeper	X	X		TF,RF	U	
Xiphorhynchus obsoletus	Striped Woodcreeper	X	X		RF	C	
Xiphorhynchus pardalotus	Chestnut-rumped Woodcreeper	X	X		TF, RF, MT	C	
Xiphorhynchus guttatus	Buff-throated Woodcreeper	X	X		RF	U	
Lepidocolaptes albolineatus	Lineated Woodcreeper	X			TF	U	
Campyloramphus procurvoides	Curve-billed Scythebill	X	X		RF, TF	U	
Thamnophilidae	**Typical Antbirds**						
Cymbilaimus lineatus	Fasciated Antshrike	X	X		RF, TF	F	
Frederickena viridis	Black-throated Antshrike	X			RF	U	
Sakesphorus canadensis	Black-crested Antshrike		X		RF	U	
Sakesphorus melanothorax	Band-tailed Antshrike	X			SG	U	
Thamnophilus murinus	Mouse-colored Antshrike	X	X		TF, RF	F	
Thamnophilus punctatus	Guianan Slaty-Antshrike		X		SF	F	
Thamnophilus amazonicus	Amazonian Antshrike	X	X		RF, TF	C	
Pygiptila stellaris	Spot-winged Antshrike	X	X		RF	F	

Scientific name	English common name	Acarai	Kamoa	Transit	Habitat	Abundance	Elevation
Dysithamnus mentalis	Plain Antvireo	X			MT	U	min elev 570 m
Thamnomanes ardesiacus	Dusky-throated Antshrike	X	X		TF, RF	C	
Thamnomanes caesius	Cinereous Antshrike	X	X		TF, RF	C	
Myrmotherula brachyura	Pygmy Antwren	X	X		RF, SG, TF	C	
Myrmotherula surinamensis	Guianan Streaked-Antwren	X	X		RF, RI	C	
Myrmotherula guttata	Rufous-bellied Antwren	X	X		RF	U	
Myrmotherula gutturalis	Brown-bellied Antwren	X	X		TF, RF	F	
Myrmotherula axillaris	White-flanked Antwren	X	X		TF, RF, SG, SF	C	
Myrmotherula longipennis	Long-winged Antwren	X	X		TF, RF	F	
Myrmotherula menetriesii	Gray Antwren	X	X		TF, RF	F	
Herpsilochmus sticturus	Spot-tailed Antwren	X	X		RF	F	
Herpsilochmus stictocephalus	Todd's Antwren	X	X		TF, MT, SF	C	
Hersilochmus rufomarginatus	Rufous-winged Antwren	X			MT	C	
Microrhopias quixensis	Dot-winged Antwren	X	X		SG, RI	C	
Terenura callinota	Rufous-rumped Antwren	X			MT	U	
Terenura spodioptila	Ash-winged Antwren	X	X		TF, RF	F	
Cercomacra cinerascens	Gray Antbird	X	X		TF, RF	F	
Cercomacra tyrannina	Dusky Antbird	X	X		RF, SG	F	
Myrmoborus leucophrys	White-browed Antbird	X			SG, RF	F	
Hypocnemis cantator	Guianan Warbling-Antbird	X	X		TF, RF, SG, MT	F	
Hypocnemoides melanopogon	Black-chinned Antbird	X	X		RI	C	
Hylophylax naevius	Spot-backed Antbird	X			TF, RF	C	
Hylophylax poecilonotus	Scale-backed Antbird	X			TF	F	
Schistocichla leucostigma	Spot-winged Antbird	X			RF	U	
Sclateria naevia	Silvered Antbird	X			SG	U	
Percnostola rufifrons	Black-headed Antbird	X	X		TF, RF, MT	F	
Myrmeciza ferruginea	Ferruginous-backed Antbird	X	X		TF, RF	F	
Pithys albifrons	White-plumed Antbird	X	X		TF, RF	F	
Gymnopithys rufigula	Rufous-throated Antbird	X	X		TF, RF	F	
Formicariidae	**Ground Antbirds**						
Formicarius colma	Rufous-capped Antthrush	X			RF	U	
Formicarius analis	Black-faced Antthrush	X			RF, TF	U	
Grallaria varia	Variegated Antpitta	X	X		TF, RF	U	
Hylopezus macularius	Spotted Antpitta	X	X		RF	U	
Myrmothera campanisona	Thrush-like Antpitta	X	X		TF, RF	F	
Tyrannidae	**Tyrant Flycatchers**						
Phyllomyias griseiceps	Sooty-headed Tyrannulet	X			TF	U	
Zimmerius gracilipes	Slender-footed Tyrannulet	X	X		TF, RF, MT, SF	C	
Ornithion inerme	White-lored Tyrannulet	X	X		RF, TF	U	
Camptostoma obsoletum	Southern Beardless-Tyrannulet	X			TF	U	
Tyrannulus elatus	Yellow-crowned Tyrannulet	X	X		RF, TF	F	
Myiopagis caniceps	Gray Elaenia	X	X		TF, RF	U	
Myiopagis gaimardii	Forest Elaenia	X	X		TF, RF	F	

Scientific name	English common name	Acarai	Kamoa	Transit	Habitat	Abundance	Elevation
Myiopagis flavivertex	Yellow-crowned Elaenia		X		RF, RI	F	
Mionectes macconnelli	McConnell's Flycatcher	X	X		TF	F	
Leptopogon amaurocephalus	Sepia-capped Flycatcher	X			TF	U	
Corythopis torquatus	Ringed Antpipit	X			RF, TF	F	
Myiornis ecaudatus	Short-tailed Pygmy-Tyrant	X	X		RF, TF	U	
Lophotriccus vitiosus	Double-banded Pygmy-Tyrant	X	X		TF, RF, SG	C	
Lophotriccus galeatus	Helmeted Pygmy-Tyrant	X	X		TF	F	
Hemitriccus josephinae	Boat-billed Tody-Tyrant	X	X		RF	U	
Hemitriccus zosterops	White-eyed Tody-Tyrant	X	X		TF	F	
Todirostrum pictum	Painted Tody-Flycatcher		X		TF	U	
Ramphotrigon megacephalum	Large-headed Flatbill	X			SG	U	
Ramphotrigon ruficauda	Rufous-tailed Flatbill	X	X		RF, SG	C	
Rhynchocyclus olivaceus	Olivaceus Flatbill	X			TF	U	
Tolmomyias assimilis	Yellow-margined Flycatcher	X	X		TF, RF	F	
Tolmomyias poliocephalus	Gray-crowned Flycatcher	X	X		RF, TF	F	
Platyrinchus saturatus	Cinnamon-crested Spadebill	X	X		TF, RF	U	
Platyrinchus coronatus	Golden-crowned Spadebill	X	X		TF, RF	U	
Myiobius barbatus	Whiskered Flycatcher	X			TF, RF	U	
Neopipo cinnamomea	Cinnamon Manakin-Tyrant	X			TF, RF	U	
Contopus nigrescens	Blackish Pewee	X			MT	U	min elev 580 m
Contopus cinereus	Tropical Pewee	X			MT	U	
Lathrotriccus euleri	Euler's Flycatcher		X		RF, RI	F	
Hirundinea ferruginea	Cliff Flycatcher	X			MT	U	
Attila spadiceus	Bright-rumped Attila	X	X		TF, RF	F	
Attila cinnamomeus	Cinnamon Attila	X	X		RF	C	
Rhytipterna simplex	Grayish Mourner	X	X		TF, RF	F	
Myiarchus tuberculifer	Dusky-capped Flycatcher	X	X		TF, RF, MT, SF	U	
Pitangus sulphuratus	Great Kiskadee		X		RF	U	
Myiozetetes luteiventris	Dusky-chested Flycatcher	X			TF	U	
Conopias albovittatus	White-ringed Flycatcher	X	X		TF, RF	U	
Legatus leucophaius	Piratic Flycatcher	X	X		RF, RI	U	
Tyrannus melancholicus	Tropical Kingbird			X	RI		
Oxyruncidae	**Sharpbills**						
Oxyruncus cristatus	Sharpbill	X			TF	U	
Cotingidae	**Cotingas**						
Cotinga cayana	Spangled Cotinga		X		RF	U	
Lipaugus vociferans	Screaming Piha	X	X		TF, RF	C	
Xipholena punicea	Pompadour Cotinga	X	X		TF, RF, SF	U	
Procnias albus	White Bellbird	X	X		TF	U	
Querula purpurata	Purple-throated Fruitcrow	X			TF	F	
Perissocephalus tricolor	Capuchinbird	X	X		TF, RF	U	
Phoenicercus carnifex	Guianan Red-Cotinga	X	X		RF, TF	F	
Rupicola rupicola	Guianan Cock-of-the-Rock	X	X		RF, MT	F	
Pipridae	**Manakins**						
Xenopipo atronitens	Black Manakin		X		SF	F	
Manacus manacus	White-bearded Manakin		X		RF	U	

Scientific name	English common name	Acarai	Kamoa	Transit	Habitat	Abundance	Elevation
Corapipo gutturalis	White-throated Manakin	X	X		TF, MT	C	min elev 380 m
Pipra erythrocephala	Golden-headed Manakin	X	X		TF, RF, SG	C	
Pipra pipra	White-crowned Manakin	X	X		TF, RF	C	
Lepidothrix serena	White-fronted Manakin	X			RF, TF	C	
Tyranneutes virescens	Tiny Tyrant-Manakin	X	X		TF, RF	C	
Neopelma chrysocephalus	Saffron-crested Tyrant-Manakin		X		SF	F	
Incertae Sedis	**Taxonomic placement uncertain**						
Schiffornis turdina	Thrush-like Schiffornis	X	X		TF, RF	F	
Piprites chloris	Wing-barred Piprites	X	X		TF, RF	U	
Laniocera hypopyrra	Cinereous Mourner		X		RF	U	
Pachyramphus marginatus	Black-capped Becard	X	X		TF, RF	F	
Vireonidae	**Vireos**						
Cyclarhis gujanensis	Rufous-browed Peppershrike	X			MT	F	
Vireolanius leucotis	Slaty-capped Shrike-Vireo	X	X		TF, RF	F	
Vireo olivaceus	Red-eyed Vireo	X			TF	U	
Hylophilus thoracicus	Lemon-chested Greenlet	X	X		RF, TF	F	
Hylophilus sclateri	Tepui Greenlet	X	X		MT	F	min elev 530 m
Hylophilus muscicapinus	Buff-cheeked Greenlet	X	X		TF, RF	C	
Hylophilus ochraceiceps	Tawny-crowned Greenlet	X			TF, RF	U	
Corvidae	**Jays**						
Cyanocorax cayanus	Cayenne Jay	X	X		SF, RI	F	
Hirundinidae	**Swallows**						
Progne tapera	Brown-chested Martin			X	RI		
Progne chalybea	Gray-breasted Martin			X	RI		
Tachycineta albiventer	White-winged Swallow			X	RI		
Atticora fasciata	White-banded Swallow		X		RI	U	
Neochelidon tibialis	White-thighed Swallow	X			MT	C	
Hirundo rustica	Barn Swallow			X	RI		
Troglodytidae	**Wrens**						
Thryothorus coraya	Coraya Wren	X	X		TF, RF, SG, MT	F	
Henicorhina leucosticta	White-breasted Wood-Wren	X			MT	F	min elev 600 m
Cyphorhinus arada	Musician Wren	X			TF, RF	U	
Microcerculus bambla	Wing-banded Wren	X			TF	U	
Sylviidae	**Gnatwrens, Gnatcatchers**						
Microbates collaris	Collared Gnatwren	X			TF	U	
Ramphocaenus melanurus	Long-billed Gnatwren	X	X		RF, TF, MT, SG	C	
Turdidae	**Thrushes**						
Turdus fumigatus	Cocoa Thrush	X	X		RF	F	
Turdus albicollis	White-necked Thrush	X	X		TF, RF	F	
Thraupidae	**Tanagers**						
Lamprospiza melanoleuca	Red-billed Pied Tanager	X			TF	U	
Hemithraupis flavicollis	Yellow-backed Tanager	X	X		TF	U	
Lanio fulvus	Fulvous Shrike-Tanager	X			TF	F	
Tachyphonus cristatus	Flame-crested Tanager	X			TF	U	
Tachyphonus surinamus	Fulvous-crested Tanager	X	X		TF, RF	F	

Scientific name	English common name	Acarai	Kamoa	Transit	Habitat	Abundance	Elevation
Ramphocelus carbo	Silver-beaked Tanager			X	RI		
Cyanicterus cyanicterus	Blue-backed Tanager	X	X		TF, RF	U	
Tangara velia	Opal-rumped Tanager	X	X		TF	U	
Tangara chilensis	Paradise Tanager		X		RF	F	
Tangara punctata	Spotted Tanager	X	X		TF, MT	F	
Tangara gyrola	Bay-headed Tanager	X	X		TF, RF	F	
Dacnis cayana	Blue Dacnis	X	X		TF, RF	U	
Dacnis lineata	Black-faced Dacnis	X			TF	U	
Chlorophanes spiza	Green Honeycreeper	X	X		TF, RF	F	
Cyanerpes caeruleus	Purple Honeycreeper	X	X		TF, RF	F	
Cyanerpes cyaneus	Red-legged Honeycreeper	X	X		TF, RF	F	
Tersina viridis	Swallow-Tanager	X	X		RI, RF	C	
Incertae Sedis	**Taxonomic placement uncertain**						
Coereba flaveola	Bananaquit	X	X		TF, RF, MT, SG, SF	C	
Emberizidae	**Emberizine Finches**						
Arremon taciturnus	Pectoral Sparrow	X			TF	U	
Paroaria gularis	Red-capped Cardinal		X		RI	U	
Cardinalidae	**Grosbeaks, Saltators**						
Saltator maximus	Buff-throated Saltator	X			RF, SG	U	
Saltator grossus	Slate-colored Grosbeak	X	X		TF, RF, SG	F	
Cyanocompsa cyanoides	Blue-black Grosbeak	X	X		TF, RF, SG	F	
Caryothraustes canadensis	Yellow-green Grosbeak	X	X		TF, RF	F	
Parulidae	**Wood Warblers**						
Parula pitiayumi	Tropical Parula	X			TF, MT	F	
Dendroica striata	Blackpoll Warbler		X		RF	U	
Setophaga ruticilla	American Redstart	X			MT	F	min elev 640 m
Phaeothlypis rivularis	River Warbler	X	X		RI, SG	F	
Incertae Sedis	**Taxonomic placement uncertain**						
Granatellus pelzelni	Rose-breasted Chat	X	X		RF, SG	F	
Icteridae	**New World Blackbirds**						
Icterus cayanensis	Epaulet Oriole	X			RF	U	
Cacicus cela	Yellow-rumped Cacique		X		RI, RF	F	
Cacicus haemorrhous	Red-rumped Cacique		X		RI, RF	F	
Psarocolius viridis	Green Oropendola	X	X		TF, RF	F	
Fringillidae	**Cardueline Finches**						
Euphonia cayennensis	Golden-sided Euphonia	X	X		TF, RF	F	
Euphonia chrysopasta	Golden-bellied Euphonia		X		RF	U	
Euphonia sp.	unidentified Euphonia	X			TF	F	
Chlorophonia cyanea	Blue-naped Chlorophonia	X			MT	F	
Species totals		**250**	**232**				

Appendix 4

Preliminary Checklist of Non-Volant Mammals of the Konashen COCA, Southern Guyana

James G. Sanderson, Eustace Alexander, Vitus Antone, and Charakura Yukuma

The presence of mammals gleaned from interviews, surveys, and camera traps from October 4 – 27, 2006 in Konashen COCA, southern Guyana. Other camera trapping efforts at Wanakoko and along the old Suriname trail are included in the column labeled **Additional**. The Red List assessments are from http://www.iucnredlist.org and include the 2008 status. Status categories are as follows:

EN = Endangered
VU = Vulnerable
NT = Near Threatened
LR = Lower Risk
LC = Least Concern
DD = Data Deficient
N/R = not ranked

Order	Family	Latin Name	Common Name	Site 1: Acarai			Site 2: Kamoa		Additional	IUCN Red List Status
				Interview	Survey	Camera trap	Survey	Camera trap	Camera trap	
Marsupiallia	Marsupialia	*Philander opossum*	Common gray four-eyed opossum	P						LR
		Metachirus nudicaudatus	Brown four-eyed opossum	P						LR
		Didelphis marsupialis	Common marsupial	P						LR
Xenarthra	Myrmecophagidae	*Myrmecophaga tridactyla*	Giant anteater	P	Photograph					NT
		Tamandua tetradactyla	Southern tamandua	P	Photograph		Photograph			LC
		Cyclopes didactylus	Silky or Pygmy anteater	P						LC
	Megalonychidae	*Bradypus tridactylus*	Pale-throated three-toed sloth	P			Photograph			LC
		Choloepus didactylus	Southern two-toed sloth	P						LC
	Dasypodidae	*Cabassous unicinctus*	Northern naked-tailed armadillo							LC
		Priodontes maximus	Giant armadillo	P					Photograph	VU
		Dasypus novemcinctus	Nine-banded armadillo							LC
		Dasypus kappleri	Great long-nosed armadillo	P					Photograph	LC
Primates	Callitrichidae	*Saguinus midas*	Golden-handed or Midas tamarin	P	Observed		Observed			LC
	Cebidae	*Saimiri sciureus*	Common squirrel monkey	P	Observed		Observed			LC
		Cebus apella	Brown capuchin monkey	P	Observed		Observed			LC
		Cebus olivaceus	Wedge-capped capuchin	P			Observed			LC
		Pithecia pithecia	Guianan saki monkey	P			Observed			LC
		Chiropotes satanas	Brown-bearded saki monkey	P	Observed		Observed			EN
		Alouatta seniculus	Red howler monkey	P	Heard daily		Heard daily			LC
		Ateles paniscus	Black spider monkey	P	Observed		Observed			LC
Carnivora	Canidae	*Cerdocyon thous*	Crab-eating fox	P (savanna)						LC
		Speothos venaticus	Bush dog	P						VU
	Procyonidae	*Procyon cancrivorus*	Crab-eating raccoon	P						LR
		Potos flavus	Kinkajou	P						LR

Order	Family	Latin Name	Common Name	Interview	Site 1: Acarai Survey	Site 1: Acarai Camera trap	Site 2: Kamoa Survey	Site 2: Kamoa Camera trap	Additional Camera trap	IUCN Red List Status
	Mustelidae	Galictis vittata	Grison							LR
		Eira barbara	Tayra	P			Observed		Photograph	LR
		Pteronura brasiliensis	Giant river otter	P	Observed		Observed			EN
	Felidae	Leopardus pardalis	Ocelot	P (skin)						LC
		Leopardus tigrinus	Oncilla or Tigrina							NT
		Leopardus weidii	Margay							LC
		Herpailurus yaguarondi	Jaguarundi	P						LC
		Puma concolor	Puma	P			Tracks	Photograph		NT
		Panthera onca	Jaguar	P	Observed			Photograph		NT
Perissodactyla	Tapiridae	Tapirus terrestris	Brazilian tapir	P	Observed	Photograph	Observed	Photograph	Photograph	VU
Artiodactyla	Tayassuidae	Tayassu tajacu	Collared peccary	P	Tracks	Photograph	Tracks			LR
		Tayassu pecari	White-lipped peccary	P	Tracks		Tracks		Photograph	LR
	Cervidae	Mazama americana	Red brocket deer	P		Photograph				DD
		Mazama gouazoubira	Gray brocket deer	P						DD
		Odocoileus virginianus	White-tailed deer	P (savanna)						LR
Rodentia	Sciuridae	Sciurus aestuans	Guianan squirrel	P	Observed					LR
	Erethizontidae	Coendou prehensilis	Brazilian porcupine	P						LR
		Coendou melanurus	Black-tailed hairy dwarf porcupine	P						N/A
	Hydrochaeridae	Hydrochaeris hydrochaeris	Capybara	P	Observed		Observed			LR
	Agoutidae	Agouti paca	Paca	P	Observed		Observed		Photograph	LR
	Dasyproctidae	Dasyprocta leoprina	Red-rumped agouti	P	Observed		Observed		Photograph	LR
		Myoprocta acouchy	Red acouchy	P						LR
	Leporidae	Sylvilagus brasiliensis	Tapiti or Brazilian rabbit	P (savanna)						LR
Total				42	17	3	18	2	7	

Additional Published Reports of the Rapid Assessment Program

SOUTH AMERICA

* Bolivia: Alto Madidi Region. Parker, T.A. III and B. Bailey (eds.). 1991. A Biological Assessment of the Alto Madidi Region and Adjacent Areas of Northwest Bolivia May 18 - June 15, 1990. RAP Working Papers 1. Conservation International, Washington, DC.

§ Bolivia: Lowland Dry Forests of Santa Cruz. Parker, T.A. III, R.B. Foster, L.H. Emmons and B. Bailey (eds.). 1993. The Lowland Dry Forests of Santa Cruz, Bolivia: A Global Conservation Priority. RAP Working Papers 4. Conservation International, Washington, DC.

§ Bolivia/Perú: Pando, Alto Madidi/Pampas del Heath. Montambault, J.R. (ed.). 2002. Informes de las evaluaciones biológicas de Pampas del Heath, Perú, Alto Madidi, Bolivia, y Pando, Bolivia. RAP Bulletin of Biological Assessment 24. Conservation International, Washington, DC.

* Bolivia: South Central Chuquisaca Schulenberg, T.S. and K. Awbrey (eds.). 1997. A Rapid Assessment of the Humid Forests of South Central Chuquisaca, Bolivia. RAP Working Papers 8. Conservation International, Washington, DC.

* Bolivia: Noel Kempff Mercado National Park. Killeen, T.J. and T.S. Schulenberg (eds.). 1998. A biological assessment of Parque Nacional Noel Kempff Mercado, Bolivia. RAP Working Papers 10. Conservation International, Washington, DC.

* Bolivia: Río Orthon Basin, Pando. Chernoff, B. and P.W. Willink (eds.). 1999. A Biological Assessment of Aquatic Ecosystems of the Upper Río Orthon Basin, Pando, Bolivia. RAP Bulletin of Biological Assessment 15. Conservation International, Washington, DC.

* Brazil: Abrolhos Bank. Dutra, G.F., G.R. Allen, T. Werner and S.A. McKenna (eds.). 2005. A Rapid Marine Biodiversity Assessment of the Abrolhos Bank, Bahia, Brazil. RAP Bulletin of Biological Assessment 38. Conservation International, Washington, DC.

* Brazil: Rio Negro and Headwaters. Willink, P.W., B. Chernoff, L.E. Alonso, J.R. Montambault and R. Lourival (eds.). 2000. A Biological Assessment of the Aquatic Ecosystems of the Pantanal, Mato Grosso do Sul, Brasil. RAP Bulletin of Biological Assessment 18. Conservation International, Washington, DC.

§ Brazil: Tumucumaque National Park. Bernard, E. (ed.). 2008. Inventários Biológicos Rápidos no Parque Nacional Montanhas do Tumucumaque, Amapá, Brasil. RAP Bulletin of Biological Assessment 48. Conservation International, Arlington, VA.

* Ecuador: Cordillera de la Costa. Parker, T.A. III and J.L. Carr (eds.). 1992. Status of Forest Remnants in the Cordillera de la Costa and Adjacent Areas of Southwestern Ecuador. RAP Working Papers 2. Conservation International, Washington, DC.

* Ecuador/Perú: Cordillera del Condor. Schulenberg, T.S. and K. Awbrey (eds.). 1997. The Cordillera del Condor of Ecuador and Peru: A Biological Assessment. RAP Working Papers 7. Conservation International, Washington, DC.

* Ecuador/Perú: Pastaza River Basin. Willink, P.W., B. Chernoff and J. McCullough (eds.). 2005. A Rapid Biological Assessment of the Aquatic Ecosystems of the Pastaza River Basin, Ecuador and Perú. RAP Bulletin of Biological Assessment 33. Conservation International, Washington, DC.

§ Guyana: Kanuku Mountain Region. Parker, T.A. III and A.B. Forsyth (eds.). 1993. A Biological Assessment of the Kanuku Mountain Region of Southwestern Guyana. RAP Working Papers 5. Conservation International, Washington, DC.

* Guyana: Eastern Kanuku Mountains. Montambault, J.R. and O. Missa (eds.). 2002. A Biodiversity Assessment of the Eastern Kanuku Mountains, Lower Kwitaro River, Guyana. RAP Bulletin of Biological Assessment 26. Conservation International, Washington, DC.

* Paraguay: Río Paraguay Basin. Chernoff, B., P.W. Willink and J. R. Montambault (eds.). 2001. A biological assessment of the Río Paraguay Basin, Alto Paraguay, Paraguay. RAP Bulletin of Biological Assessment 19. Conservation International, Washington, DC.

* Perú: Tambopata-Candamo Reserved Zone. Foster, R.B., J.L. Carr and A.B. Forsyth (eds.). 1994. The Tambopata-Candamo Reserved Zone of southeastern Perú: A Biological Assessment. RAP Working Papers 6. Conservation International, Washington, DC.

* Perú: Cordillera de Vilcabamba. Alonso, L.E., A. Alonso, T. S. Schulenberg and F. Dallmeier (eds.). 2001. Biological and Social Assessments of the Cordillera de Vilcabamba, Peru. RAP Working Papers 12 and SI/MAB Series 6. Conservation International, Washington, DC.

* Suriname: Coppename River Basin. Alonso, L.E. and H.J. Berrenstein (eds.). 2006. A rapid biological assessment of the aquatic ecosystems of the Coppename River Basin, Suriname. RAP Bulletin of Biological Assessment 39. Conservation International, Washington, DC.

* Suriname: Lely and Nassau Plateaus. Alonso, L.E. and J.H. Mol (eds.). 2007. A Rapid Biological Assessment of the Lely and Nassau Plateaus, Suriname (with additional information on the Brownsberg Plateau). RAP Bulletin of Biological Assessment 43. Conservation International, Arlington, VA.

* Venezuela: Caura River Basin. Chernoff, B., A. Machado-Allison, K. Riseng and J.R. Montambault (eds.). 2003. A Biological Assessment of the Aquatic Ecosystems of the Caura River Basin, Bolívar State, Venezuela. RAP Bulletin of Biological Assessment 28. Conservation International, Washington, DC.

* Venezuela: Orinoco Delta and Gulf of Paria. Lasso, C.A., L.E. Alonso, A.L. Flores and G. Love (eds.). 2004. Rapid assessment of the biodiversity and social aspects of the aquatic ecosystems of the Orinoco Delta and the Gulf of Paria, Venezuela. RAP Bulletin of Biological Assessment 37. Conservation International, Washington, DC.

* Venezuela: Rio Paragua. Señaris, J.C., C.A. Lasso and A.L. Flores (eds.). 2008. Evaluación Rapida de la Biodiversidad de los Ecosistemas Acuáticos de la Cuenca Alta del Río Paragua, Estado Bolívar, Venezuela. RAP Bulletin of Biological Assessment 49. Conservation International, Arlington, VA.

* Venezuela: Ventuari and Orinoco Rivers. Lasso, C.A., J.C. Señaris, L.E. Alonso, and A.L. Flores (eds.). 2006. Evaluación Rápida de la Biodiversidad de los Ecosistemas Acuáticos en la Confluencia de los ríos Orinoco y Ventuari, Estado Amazonas (Venezuela). Boletín RAP de Evaluación Biológica 30. Conservation International, Washington, DC.

CENTRAL AMERICA

§ Belize: Columbia River Forest Reserve. Parker, T.A. III. (ed.). 1993. A Biological Assessment of the Columbia River Forest Reserve, Toledo District, Belize. RAP Working Papers 3. Conservation International, Washington, DC.

* Guatemala: Laguna del Tigre National Park. Bestelmeyer, B. and L.E. Alonso (eds.). 2000. A Biological Assessment of Laguna del Tigre National Park, Petén, Guatemala. RAP Bulletin of Biological Assessment 16. Conservation International, Washington, DC.